Hollow Earth Collection

by
WILLIS GEORGE and ADAM SEABORN

Hollow Earth Collection

© 2021 Joy Garcia

This information represents work that is found and part of the public domain. This publication may be reproduced, stored in a retrieval system, and/or transmitted by means of electronic, mechanical, photocopy, recording or other medium especially for educational purposes as it is part of the public domain. This book is designed to provide accurate and authoritative information with regard to the subject matter covered. Compiled, formatted, edited, and produced by Joy Garcia for Sacred Word Publishing LLC for public consideration.

Sacred Word Publishing

www.sacredwordpublishing.com

SacredWordPublishingLLC@gmail.com

Published by Sacred Word Publishing, LLC.

for public consideration.

1st Printing: 2021

978-1-716-08632-8

The Smoky God

Or

A Voyage to the Inner World

"He is the God who sits in the center, on the navel of the earth, and he is the interpreter of religion to all mankind."—Plato.

PART ONE

AUTHOR'S FOREWORD

I fear the seemingly incredible story which I am about to relate will be regarded as the result of a distorted intellect superinduced, possibly, by the glamour of unveiling a marvelous mystery, rather than a truthful record of the unparalleled experiences related by one Olaf Jansen, whose eloquent madness so appealed to my imagination that all thought of an analytical criticism has been effectually dispelled.

Marco Polo will doubtless shift uneasily in his grave at the strange story I am called upon to chronicle, a story as strange as a Munchausen tale. It is also incongruous that I, a disbeliever, should be the one to edit the story of Olaf Jansen, whose name is now for the first time given to the world, yet who must hereafter rank as one of the notables of earth.

I freely confess his statements admit of no rational analysis but have to do with the profound mystery concerning the frozen North that for centuries has claimed the attention of scientists and laymen alike.

However much they are at variance with the cosmographical manuscripts of the past, these plain statements may be relied upon as a record of the things Olaf Jansen claims to have seen with his own eyes.

A hundred times I have asked myself whether it is possible that the world's geography is incomplete, and that the startling narrative of Olaf Jansen is predicated upon demonstrable facts. The reader may be able to answer these queries to his own satisfaction, however far the chronicler of this narrative may be from having reached a conviction. Yet sometimes even I am at a loss to know whether I have been led away from an abstract truth by the *ignes fatui* of a clever superstition, or whether heretofore accepted facts are, after all, founded upon falsity.

It may be that the true home of Apollo was not at Delphi, but in that older earth-center of which Plato speaks, where he says: "Apollo's real home is among the Hyperboreans, in a land of perpetual life, where mythology tells us two doves flying from the two opposite ends of the world met in this fair region, the home of Apollo. Indeed, according to Hecatæus, Leto, the mother of Apollo, was born on an island in the Arctic Ocean far beyond the North Wind."

It is not my intention to attempt a discussion of the theogony of the deities nor the cosmogony of the world. My simple duty is to enlighten the world concerning a heretofore unknown portion of the universe, as it was seen and described by the old Norseman, Olaf Jansen.

Interest in northern research is international. Eleven nations are engaged in, or have contributed to, the perilous work of trying to solve Earth's one remaining cosmological mystery.

There is a saying, ancient as the hills, that "truth is stranger than fiction," and in a most startling manner has this axiom been brought home to me within the last fortnight.

It was just two o'clock in the morning when I was aroused from a restful sleep by the vigorous ringing of my doorbell. The untimely disturber proved to be a messenger bearing a note, scrawled almost to the point of illegibility, from an old Norseman by the name of Olaf Jansen. After much deciphering, I made out the writing, which simply said: "Am ill unto death. Come." The call was imperative, and I lost no time in making ready to comply.

Perhaps I may as well explain here that Olaf Jansen, a man who quite recently celebrated his ninety-fifth birthday, has for the last half-dozen years been living alone in an unpretentious bungalow out Glendale way, a short distance from the business district of Los Angeles, California.

It was less than two years ago, while out walking one afternoon, that I was attracted by Olaf Jansen's house and its homelike surroundings, toward its owner and occupant, whom I afterward came to know as a believer in the ancient worship of Odin and Thor.

There was a gentleness in his face, and a kindly expression in the keenly alert gray eyes of this man who had lived more than four-score years and ten; and, withal, a sense of loneliness that appealed to my sympathy. Slightly stooped, and with his hands clasped behind him, he walked back and forth with slow and measured tread, that day when first we met. I can hardly say what particular motive impelled me to pause in my walk and engage him in conversation. He seemed pleased when I complimented him on the attractiveness of his bungalow, and on the well-tended vines and flowers clustering in profusion over its windows, roof and wide piazza.

I soon discovered that my new acquaintance was no ordinary person, but one profound and learned to a remarkable degree; a man who, in the later years of his long life, had dug deeply into books and become strong in the power of meditative silence.

I encouraged him to talk, and soon gathered that he had resided only six or seven years in Southern California but had passed the dozen years prior in one of the middle Eastern states. Before that he had been a fisherman off the coast of Norway, in the region of the Lofoden Islands, from whence he had made trips still farther north to Spitzbergen and even to Franz Josef Land.

When I started to take my leave, he seemed reluctant to have me go, and asked me to come again. Although at the time I thought nothing of it, I remember now that he made a peculiar remark as I extended my hand in leave-taking. "You will come again?" he asked. "Yes, you will come again someday. I am sure you will; and I shall show you my library and tell you many things of which you have never dreamed, things so wonderful that it maybe you will not believe me."

I laughingly assured him that I would not only come again but would be ready to believe whatever he might choose to tell me of his travels and adventures.

In the days that followed I became well acquainted with Olaf Jansen, and, little by little, he told me his story, so marvelous, that its very daring challenges reason and belief. The old Norseman always expressed himself with so much earnestness and sincerity that I became enthralled by his strange narrations.

Then came the messenger's call that night, and within the hour I was at Olaf Jansen's bungalow.

He was very impatient at the long wait, although after being summoned I had come immediately to his bedside.

"I must hasten," he exclaimed, while yet he held my hand in greeting.

"I have much to tell you that you know not, and I will trust no one but you. I fully realize," he went on hurriedly, "that I shall not survive the night. The time has come to join my fathers in the great sleep."

I adjusted the pillows to make him more comfortable and assured him I was glad to be able to serve him in any way possible, for I was beginning to realize the seriousness of his condition.

The lateness of the hour, the stillness of the surroundings, the uncanny feeling of being alone with the dying man, together with his weird story, all combined to make my heart beat fast and loud with a feeling for which I have no name. Indeed, there were many times that night by the old Norseman's couch, and there have been many times since, when a sensation rather than a conviction took possession of my very soul, and I seemed not only to believe in, but actually see, the strange lands, the strange people and the strange world of which he told, and to hear the mighty orchestral chorus of a thousand lusty voices.

For over two hours he seemed endowed with almost superhuman strength, talking rapidly, and to all appearances, rationally. Finally, he gave into my hands certain data, drawings and crude maps. "These," said he in conclusion, "I leave in your hands. If I can have your promise to give them to the world, I shall die happy, because I desire that people may know the truth, for then all mystery concerning the frozen Northland will be explained. There is no chance of your suffering the fate I suffered. They will not put you in irons, nor confine you in a mad-house, because you are not telling your own story, but mine, and I, thanks to the gods, Odin and Thor, will be in my grave, and so beyond the reach of disbelievers who would persecute."

Without a thought of the far-reaching results the promise entailed or foreseeing the many sleepless nights which the obligation has since brought me, I gave my hand and with it a pledge to discharge faithfully his dying wish.

As the sun rose over the peaks of the San Jacinto, far to the eastward, the spirit of Olaf Jansen, the navigator, the explorer and worshiper of Odin and Thor, the man whose experiences and travels, as related, are without a parallel in all the world's history, passed away, and I was left alone with the dead.

And now, after having paid the last sad rites to this strange man from the Lofoden Islands, and the still farther "Northward Ho!", the courageous explorer of frozen regions, who in his declining years (after he had passed the four-score mark) had sought an asylum of restful peace in sun-favored California, I will undertake to make public his story.

But first of all, let me indulge in one or two reflections:

Generation follows generation, and the traditions from the misty past are handed down from sire to son, but for some strange reason interest in the ice-locked unknown does not abate with the receding years, either in the minds of the ignorant or the tutored.

With each new generation a restless impulse stirs the hearts of men to capture the veiled citadel of the Arctic, the circle of silence, the land of glaciers, cold wastes of waters and winds that are strangely warm. Increasing interest is manifested in the mountainous icebergs, and marvelous speculations are indulged in concerning the earth's center of gravity, the cradle of the tides, where the whales have their nurseries, where the magnetic needle goes mad, where the Aurora Borealis illumines the night, and where brave and courageous spirits of every generation dare to venture and explore, defying the dangers of the "Farthest North."

One of the ablest works of recent years is "Paradise Found, or the Cradle of The Human Race at the North Pole," by William F. Warren. In his carefully prepared volume, Mr. Warren almost stubbed his toe against the real truth, but missed it seemingly by only a hair's breadth, if the old Norseman's revelation be true.

Dr. Orville Livingston Leech, scientist, in a recent article, says:

"The possibilities of a land inside the earth were first brought to my attention when I picked up a geode on the shores of the Great Lakes. The geode is a spherical and apparently solid stone, but when broken is found to be hollow and coated with crystals. The earth is only a larger form of a geode, and the law that created the geode in its hollow form undoubtedly fashioned the earth in the same way."

In presenting the theme of this almost incredible story, as told by Olaf Jansen, and supplemented by manuscript, maps and crude drawings entrusted to me, a fitting introduction is found in the following quotation:

"In the beginning God created the heaven and the earth, and the earth was without form and void." And also, "God created man in his own image." Therefore, even in things material, man must be God-like, because he is created in the likeness of the Father.

A man builds a house for himself and family. The porches or verandas are all without and are secondary. The building is really constructed for the conveniences within.

Olaf Jansen makes the startling announcement through me, an humble instrument, that in like manner, God created the earth for the "within"—that is to say, for its lands, seas, rivers, mountains, forests and valleys, and for its other internal conveniences, while the outside surface of the earth is merely the veranda, the porch, where things grow by comparison but sparsely, like the lichen on the mountain side, clinging determinedly for bare existence.

Take an eggshell, and from each end break out a piece as large as the end of this pencil. Extract its contents, and then you will have a perfect representation of Olaf Jansen's earth. The distance from the inside surface to the outside surface, according to him, is about three hundred miles. The center of gravity is not in the center of the earth, but in the center of the shell or crust; therefore, if the thickness of the earth's crust or shell is three hundred miles, the center of gravity is one hundred and fifty miles below the surface.

In their log-books Arctic explorers tell us of the dipping of the needle as the vessel sails in regions of the farthest north known. In reality, they are at the curve; on the edge of the shell, where gravity is geometrically increased, and while the electric current seemingly dashes off into space toward the phantom idea of the North Pole, yet this same electric current drops again and continues its course southward along the inside surface of the earth's crust.

In the appendix to his work, Captain Sabine gives an account of experiments to determine the acceleration of the pendulum in different latitudes. This appears to have resulted from the joint labor of Peary and Sabine. He says: "The accidental discovery that a pendulum on being removed from Paris to the neighborhood of the equator increased its time of vibration, gave the first step to our present knowledge that the polar axis of the globe is less than the equatorial; that the force of gravity at the surface of the earth increases progressively from the equator toward the poles."

According to Olaf Jansen, in the beginning this old world of ours was created solely for the "within" world, where are located the four great rivers—the Euphrates, the Pison, the Gihon and the Hiddekel. These same names of rivers, when applied to streams on the "outside" sur- face of the earth, are purely traditional from an antiquity beyond the memory of man.

On the top of a high mountain, near the fountain-head of these four rivers, Olaf Jansen, the Norseman, claims to have discovered the long-lost "Garden of Eden," the veritable navel of the earth, and to have spent over two years studying and reconnoitering in this marvelous "within" land, exuberant with stupendous plant life and abounding in giant animals; a land where the people live to be centuries old, after the order of Methuselah and other Biblical characters; a region where one-quarter of the "inner" surface is water and three-quarters land; where there are large oceans and many rivers and lakes; where the cities are superlative in construction and magnificence; where modes of transportation are as far in advance of ours as we with our boasted achievements are in advance of the inhabitants of "darkest Africa."

The distance directly across the space from inner surface to inner surface is about six hundred miles less than the recognized diameter of the earth. In the identical center of this vast vacuum is the seat of electricity—a mammoth ball of dull red fire—not startlingly brilliant, but surrounded by a white, mild, luminous cloud, giving out uniform warmth, and held in its place in the center of this internal space by the immutable law of gravitation. This electrical cloud is known to the people "within" as the abode of "The Smoky God." They believe it to be the throne of "The Most High."

Olaf Jansen reminded me of how, in the old college days, we were all familiar with the laboratory demonstrations of centrifugal motion, which clearly proved that, if the earth were a solid, the rapidity of its revolution upon its axis would tear it into a thousand fragments.

The old Norseman also maintained that from the farthest points of land on the islands of Spitzbergen and Franz Josef Land, flocks of geese may be seen annually flying still farther northward, just as the sailors and explorers record in their logbooks. No scientist has yet been audacious enough to attempt to explain, even to his own satisfaction, toward what lands these winged fowls are guided by their subtle instinct. However, Olaf Jansen has given us a most reasonable explanation.

The presence of the open sea in the Northland is also explained. Olaf Jansen claims that the northern aperture, intake or hole, so to speak, is about fourteen hundred miles across. In connection with this, let us read what Explorer Nansen writes, on page 288 of his book: "I have never had such a splendid sail. On to the north, steadily north, with a good wind, as fast as steam and sail can take us, an open sea mile after mile, watch after watch, through these unknown regions, always clearer and clearer of ice, one might almost say: 'How long will it last 2' The eye always turns to the northward as one paces the bridge. It is gazing into the future. But there is always the same dark sky ahead which means open sea." Again, the Norwood Review of England, in its issue of May 10, 1884, says: "We do not admit that there is ice up to the Pole—once inside the great ice barrier, a new world breaks upon the explorer, the climate is mild like that of England, and, afterward, balmy as the Greek Isles."

Some of the rivers "within," Olaf Jansen claims, are larger than our Mississippi and Amazon rivers combined, in point of volume of water carried; indeed their greatness is occasioned by their width and depth rather than their length, and it is at the mouths of these mighty rivers, as they flow northward and southward along the inside surface of the earth, that mammoth icebergs are found, some of them fifteen and twenty miles wide and from forty to one hundred miles in length.

Is it not strange that there has never been an iceberg encountered either in the Arctic or Antarctic Ocean that is not composed of fresh water? Modern scientists claim that freezing eliminates the salt, but Olaf Jansen claims differently.

Ancient Hindoo, Japanese and Chinese writings, as well as the hieroglyphics of the extinct races of the North American continent, all speak of the custom of sun-worshiping, and it is possible, in the startling light of Olaf Jansen's revelations, that the people of the inner world, lured away by glimpses of the sun as it shone upon the inner surface of the earth, either from the northern or the southern opening, became dissatisfied with "The Smoky God," the great pillar or mother cloud of electricity, and, weary of their continuously mild and pleasant atmosphere, followed the brighter light, and were finally led beyond the ice belt and scattered over the "outer" surface of the earth, through Asia, Europe, North America and, later, Africa, Australia and South America.

It is a notable fact that, as we approach the Equator, the stature of the human race grows less. But the Patagonians of South America are probably the only aborigines from the center of the earth who came out through the aperture usually designated as the South Pole, and they are called the giant race.

Olaf Jansen avers that, in the beginning, the world was created by the Great Architect of the Universe, so that man might dwell upon its "inside" surface, which has ever since been the habitation of the "chosen."

They who were driven out of the "Garden of Eden" brought their traditional history with them.

The history of the people living "within" contains a narrative suggesting the story of Noah and the ark with which we are familiar. He sailed away, as did Columbus, from a certain port, to a strange land he had heard of far to the northward, carrying with him all manner of beasts of the fields and fowls of the air, but was never heard of afterward.

On the northern boundaries of Alaska, and still more frequently on the Siberian coast, are found boneyards containing tusks of ivory in quantities so great as to suggest the burying-places of antiquity. From Olaf Jansen's account, they have come from the great prolific animal life that abounds in the fields and forests and on the banks of numerous rivers of the Inner World. The materials were caught in the ocean currents, or were carried on icefloes, and have accumulated like driftwood on the Siberian coast. This has been going on for ages, and hence these mysterious boneyards.

On this subject William F. Warren, in his book already cited, pages 297 and 298, says: "The Arctic rocks tell of a lost Atlantis more wonderful than Plato's. The fossil ivory beds of Siberia excel everything of the kind in the world. From the days of Pliny, at least, they have constantly been undergoing exploitation, and still they are the chief headquarters of supply. The remains of mammoths are so abundant that, as Gratacap says, 'the northern islands of Siberia seem built up of crowded bones.' Another scientific writer, speaking of the islands of New Siberia, northward of the mouth of the River Lena, uses this language: 'Large quantities of ivory are dug out of the ground every year. Indeed, some of the islands are believed to be nothing but an accumulation of drift-timber and the bodies of mammoths and other antediluvian animals frozen together.' From this we may infer that, during the years that have elapsed since the Russian conquest of Siberia, useful tusks from more than twenty thousand mammoths have been collected."

But now for the story of Olaf Jansen. I give it in detail, as set down by himself in manuscript, and woven into the tale, just as he placed them, are certain quotations from recent works on Arctic exploration, showing how carefully the old Norseman compared with his own experiences those of other voyagers to the frozen North. Thus, wrote the disciple of Odin and Thor:

PART TWO

OLAF JANSEN'S STORY

My name is Olaf Jansen. I am a Norwegian, although I was born in the little seafaring Russian town of Uleaborg, on the eastern coast of the Gulf of Bothnia, the northern arm of the Baltic Sea.

My parents were on a fishing cruise in the Gulf of Bothnia and put into this Russian town of Uleaborg at the time of my birth, being the twenty-seventh day of October 1811.

My father, Jens Jansen, was born at Rodwig on the Scandinavian coast, near the Lofoden Islands, but after marrying made his home at Stockholm, because my mother's people resided in that city. When seven years old, I began going with my father on his fishing trips along the Scandinavian coast.

Early in life I displayed an aptitude for books, and at the age of nine years was placed in a private school in Stockholm, remaining there until I was fourteen. After this I made regular trips with my father on all his fishing voyages.

My father was a man fully six feet three in height, and weighed over fifteen stone, a typical Norseman of the most rugged sort, and capable of more endurance than any other man I have ever known. He possessed the gentleness of a woman in tender little ways, yet his determination and will-power were beyond description. His will admitted of no defeat.

I was in my nineteenth year when we started on what proved to be our last trip as fishermen, and which resulted in the strange story that shall be given to the world, —but not until I have finished my earthly pilgrimage.

I dare not allow the facts as I know them to be published while I am living, for fear of further humiliation, confinement and suffering. First of all, I was put in irons by the captain of the whaling vessel that rescued me, for no other reason than that I told the truth about the marvelous discoveries made by my father and myself. But this was far from being the end of my tortures.

After four years and eight months' absence I reached Stockholm, only to find my mother had died the previous year, and the property left by my parents in the possession of my mother's people, but it was at once made over to me.

All might have been well, had I erased from my memory the story of our adventure and of my father's terrible death.

Finally, one day I told the story in detail to my uncle, Gustaf Osterlind, a man of considerable property, and urged him to fit out an expedition for me to make another voyage to the strange land.

At first, I thought he favored my project. He seemed interested and invited me to go before certain officials and explain to them, as I had to him, the story of our travels and discoveries. Imagine my disappointment and horror when, upon the conclusion of my narrative, certain papers were signed by my uncle, and, without warning, I found myself arrested and hurried away to dismal and fearful confinement in a madhouse, where I remained for twenty-eight years—long, tedious, frightful years of suffering!

"Twenty-eight years—long, tedious, frightful years of suffering."

I never ceased to assert my sanity, and to protest against the injustice of my confinement. Finally, on the seventeenth of October 1862, I was released. My uncle was dead, and the friends of my youth were now strangers. Indeed, a man over fifty years old, whose only known record is that of a madman, has no friends.

I was at a loss to know what to do for a living, but instinctively turned toward the harbor where fishing boats in great numbers were anchored, and within a week I had shipped with a fisherman by the name of Yan Hansen, who was starting on a long fishing cruise to the Lofoden Islands.

Here my earlier years of training proved of the very greatest advantage, especially in enabling me to make myself useful. This was but the beginning of other trips, and by frugal economy I was, in a few years, able to own a fishing-brig of my own.

For twenty-seven years thereafter I followed the sea as a fisherman, five years working for others, and the last twenty-two for myself.

During all these years I was a most diligent student of books, as well as a hard worker at my business, but I took great care not to mention to anyone the story concerning the discoveries made by my father and myself. Even at this late day I would be fearful of having anyone see or know the things I am writing, and the records and maps I have in my keeping. When my days on earth are finished, I shall leave maps and records that will enlighten and, I hope, benefit mankind.

The memory of my long confinement with maniacs, and all the horrible anguish and sufferings are too vivid to warrant my taking further chances.

In 1889 I sold out my fishing boats and found I had accumulated a fortune quite sufficient to keep me the remainder of my life. I then came to America.

For a dozen years my home was in Illinois, near Batavia, where I gathered most of the books in my present library, though I brought many choice volumes from Stockholm. Later, I came to

Los Angeles, arriving here March 4, 1901. The date I well remember, as it was President McKinley's second inauguration day. I bought this humble home and determined, here in the privacy of my Own abode, sheltered by my own vine and fig-tree, and with my books about me, to make maps and drawings of the new lands we had discovered, and also to write the story in detail from the time my father and I left Stockholm until the tragic event that parted us in the Antarctic Ocean.

I well remember that we left Stockholm in our fishing-sloop on the third day of April 1829, and sailed to the southward, leaving Gothland Island to the left and Oeland Island to the right. A few days later we succeeded in doubling Sandhommar Point and made our way through the sound which separates Denmark from the Scandinavian coast. In due time we put in at the town of Christians and, where we rested two days, and then started around the Scandinavian coast to the westward, bound for the Lofoden Islands.

My father was in high spirit, because of the excellent and gratifying returns he had received from our last catch by marketing at Stockholm, instead of selling at one of the seafaring towns along the Scandinavian coast. He was especially pleased with the sale of some ivory tusks that he had found on the west coast of Franz Joseph Land during one of his northern cruises the previous year, and he expressed the hope that this time we might again be fortunate enough to load our little fishing-sloop with ivory, instead of cod, herring, mackerel and salmon.

We put in at Hammerfest, latitude seventy-one degrees and forty minutes, for a few days' rest. Here we remained one week, laying in an extra supply of provisions and several casks of drinking-water, and then sailed toward Spitzbergen.

"A vessel larger than our little fishing sloop could not have threaded its way among the icebergs."

For the first few days we had an open sea and a favoring wind, and then we encountered much ice and many icebergs. A vessel larger than our little fishing-sloop could not possibly have threaded its way among the labyrinth of icebergs or squeezed through the barely open channels. These monster bergs presented an endless succession of crystal palaces, of massive cathedrals and fantastic mountain ranges, grim and sentinel-like, immovable as some towering cliff of solid rock, standing silent as a sphinx, resisting the restless waves of a fretful sea.

After many narrow escapes, we arrived at Spitzbergen on the 23d of June, and anchored at Wijade Bay for a short time, where we were quite successful in our catches. We then lifted anchor and sailed through the Hinlopen Strait and coasted along the North-East-Land. [1]

A strong wind came up from the southwest, and my father said that we had better take advantage of it and try to reach Franz Josef Land where the year before he had, by accident, found the ivory tusks that had brought him such a good price at Stockholm.

Never, before or since, have I seen so many seafowl; they were so numerous that they hid the rocks on the coast line and darkened the sky.

For several days we sailed along the rocky coast of Franz Josef Land. Finally, a favoring wind came up that enabled us to make the West Coast, and, after sailing twenty-four hours, we came to a beautiful inlet.

One could hardly believe it was the far Northland. The place was green with growing vegetation, and while the area did not comprise more than one or two acres, yet the air was warm and tranquil. It seemed to be at that point where the Gulf Stream's influence is most keenly felt. [2]

On the east coast there were numerous icebergs, yet here we were in open water. Far to the west of us, however, were icepacks, and still farther to the westward the ice appeared like ranges of low hills. In front of us, and directly to the north, lay an open sea. [3]

My father was an ardent believer in Odin and Thor and had frequently told me they were gods who came from far beyond the "North Wind."

There was a tradition, my father explained, that still farther northward was a land more beautiful than any that mortal man had ever known, and that it was inhabited by the "Chosen." 1

My youthful imagination was fired by the ardor, zeal and religious fervor of, my good father, and I exclaimed: "Why not sail to this goodly land? The sky is fair, the wind favorable and the sea open."

Even now I can see the expression of pleasurable surprise on his countenance as he turned toward me and asked: "My son, are you willing to go with me and explore—to go far beyond where man has ever ventured?" I answered affirmatively. "Very well," he replied. "May the god Odin protect us!" and, quickly adjusting the sails, he glanced at our compass, turned the prow in due northerly direction through an open channel, and our voyage had begun. 1

The sun was low in the horizon, as it was still the early summer. Indeed, we had almost four months of day ahead of us before the frozen night could come on again.

Our little fishing-sloop sprang forward as if eager as ourselves for adventure. Within thirty-six hours we were out of sight of the highest point on the coastline of Franz Josef Land. We seemed to be in a strong current running north by northeast. Far to the right and to the left of us were icebergs, but our little sloop bore down on the narrows and passed through channels and out into open seas—channels so narrow in place that, had our craft been other than small, we never could have gotten through.

On the third day we came to an island. Its shores were washed by an open sea. My father determined to land and explore for a day. This new land was destitute of timber, but we found a large accumulation of driftwood on the northern shore. Some of the trunks of the trees were forty feet long and two feet in diameter. [1]

After one day's exploration of the coastline of this island, we lifted anchor and turned our prow to the north in an open sea. [2]

I remember that neither my father nor myself had tasted food for almost thirty hours. Perhaps this was because of the tension of excitement about our strange voyage in waters farther north, my father said, than anyone had ever before been. Active mentality had dulled the demands of the physical needs.

Instead of the cold being intense as we had anticipated, it was really warmer and more pleasant than it had been while in Hammerfest on the north coast of Norway, some six weeks before. [3]

We both frankly admitted that we were very hungry, and forthwith I prepared a substantial meal from our well-stored larder. When we had partaken heartily of the repast, I told my father I believed I would sleep, as I was beginning to feel quite drowsy. "Very well," he replied, "I will keep the watch."

I have no way to determine how long I slept; I only know that I was rudely awakened by a terrible commotion of the sloop. To my surprise, I found my father sleeping soundly. I cried out lustily to him, and starting up, he sprang quickly to his feet. Indeed, had he not instantly clutched the rail; he would certainly have been thrown into the seething waves.

A fierce snowstorm was raging. The wind was directly astern, driving our sloop at a terrific speed, and was threatening every moment to capsize us. There was no time to lose, the sails had to be lowered immediately. Our boat was writhing in convulsions. A few icebergs we knew were on either side of us, but fortunately the channel was open directly to the north. But would it remain so? In front of us, girding the horizon from left to right, was a vaporish fog or mist, black as Egyptian night at the water's edge, and white like a steam-cloud toward the top, which was finally lost to view as it blended with the great white flakes of falling snow. Whether it covered a treacherous iceberg, or some other hidden obstacle against which our little sloop would dash and send us to a watery grave, or was merely the phenomenon of an Arctic fog, there was no way to determine. 1

"By what miracle we escaped being dashed to destruction, I do not know."

By what miracle we escaped being dashed to utter destruction, I do not know. I remember our little craft creaked and groaned, as if its joints were breaking. It rocked and staggered to and fro as if clutched by some fierce undertow of whirlpool or maelstrom.

Fortunately, our compass had been fastened with long screws to a crossbeam. Most of our provisions, however, were tumbled out and swept away from the deck of the cuddy, and had we not taken the precaution at the very beginning to tie ourselves firmly to the masts of the sloop, we should have been swept into the lashing sea.

Above the deafening tumult of the raging waves, I heard my father's voice. "Be courageous, my son," he shouted, "Odin is the god of the waters, the companion of the brave, and he is with us. Fear not."

To me it seemed there was no possibility of our escaping a horrible death. The little sloop was shipping water, the snow was falling so fast as to be blinding, and the waves were tumbling over our counters in reckless, white-sprayed fury. There was no telling what instant we should be dashed against some drifting icepack.

The tremendous swells would have us up to the very peaks of mountainous waves, then plunge us down into the depths of the sea's trough as if our fishing-sloop were a fragile shell. Gigantic white-capped waves, like veritable walls, fenced us in, fore and aft.

This terrible nerve-racking ordeal, with its nameless horrors of suspense and agony of fear indescribable, continued for more than three hours, and all the time we were being driven forward at fierce speed. Then suddenly, as if growing weary of its frantic exertions, the wind began to lessen its fury and by degrees to die down.

At last, we were in a perfect calm. The fog mist had also disappeared, and before us lay an iceless channel perhaps ten or fifteen miles wide, with a few icebergs far away to our right, and an intermittent archipelago of smaller ones to the left.

I watched my father closely, determined to remain silent until he spoke. Presently he untied the rope from his waist and, without saying a word, began working the pumps, which fortunately were not damaged, relieving the sloop of the water it had shipped in the madness of the storm.

He put up the sloop's sails as calmly as if casting a fishing-net, and then remarked that we were ready for a favoring wind when it came. His courage and persistence were truly remarkable.

On investigation we found less than one-third of our provisions remaining, while to our utter dismay, we discovered that our water-casks had been swept overboard during the violent plungings of our boat.

Two of our water-casks were in the main hold, but both were empty. We had â fair supply of food, but no fresh water. I realized at once the awfulness of our position. Presently I was seized with a consuming thirst. "It is indeed bad," remarked my father. "However, let us dry our bedraggled clothing, for we are soaked to the skin. Trust to the god Odin, my son. Do not give up hope."

The sun was beating down slantingly, as if we were in a southern latitude, instead of in the far Northland. It was swinging around, its orbit ever visible and rising higher and higher each day, frequently mist-covered, yet always peering through the lacework of clouds like some fretful eye of fate, guarding the mysterious Northland and jealously watching the pranks of man. Far to our right the rays decking the prisms of icebergs were gorgeous. Their reflections emitted flashes of garnet, of diamond, of sapphire. A pyrotechnic panorama of countless colors and shapes, while below could be seen the green-tinted sea, and above, the purple sky.

Footnotes

22:1 *It will be remembered that Andree started on his fatal balloon voyage from the northwest coast of Spitzbergen.*

22:2 *Sir John Barrow, Bart., F.R.S., in his work entitled "Voyages of Discovery and Research Within the Arctic Regions," says on page 57: "Mr. Beechey refers to what has frequently been found and noticed—the mildness of the temperature on the western coast of Spitzbergen, there being little or no sensation of cold, though the thermometer might be only a few degrees above the freezing-point. The brilliant and lively effect of a clear day, when the sun shines forth with a pure sky, whose azure hue is so intense as to find no parallel even in the boasted Italian sky."*

22:3 *Captain Kane, on page 299, quoting from Morton's Journal on Monday, the 26th of December, says: "As far as I could see, the open passages were fifteen miles or more wide, with sometimes mashed ice separating them. But it is all small ice, and I think it either drives out to the open space to the north or rots and sinks, as I could see none ahead to the north."*

23:1 *We find the following in "Deutsche Mythologie," page 778, from the pen of Jakob Grimm; "Then the sons of Bor built in the middle of the universe the city called Asgard, where dwell the gods and their kindred, and from that abode work out so many wondrous things both on the earth and in the heavens above it. There is in that city a place called Hlidskjalf, and when Odin is seated there upon his lofty throne he sees over the whole world and discerns all the actions of men."*

23:2 *Hall writes, on page 288: "On the 23rd of January the two Esquimaux, accompanied by two of the sea men,* p. 66 *went to Cape Lupton. They reported a sea of open water extending as far as the eye could reach."*

24:1 *Greely tells us in vol. 1, page* 100, *that:* "Privates Connell and Frederick found a large coniferous tree on the beach, just above the extreme high-water mark. It was nearly thirty inches in circumference, some thirty feet long, and had apparently been carried to that point by a current p. 68 within a couple of years. A portion of it was cut up for fire-wood, and for the first time in that valley, a bright, cheery camp-fire gave comfort to man."

24:2 Dr. Kane says, on page 379 of his works: "I cannot imagine what becomes of the ice. A strong current sets in constantly to the north; but, from altitudes of more than five hundred feet, I saw only narrow strips of ice, with great spaces of open water, from ten to fifteen miles in breadth, between them. It must, therefore, either go to an open space in the north, or dissolve."

24:3 Captain Peary's second voyage relates another circumstance which may serve to confirm a conjecture which has long been maintained by some, that an open sea, free of ice, exists at or near the Pole. "On the second of November," says Peary, "the wind freshened up to a gale from north by p. 70 west, lowered the thermometer before midnight to 5 degrees, whereas a rise of wind at Melville Island was generally accompanied by a simultaneous rise in the thermometer at low temperatures. May not this," he asks, "be occasioned by the wind blowing over an open sea in the quarter from which the wind blows? And tend to confirm the opinion that at or near the Pole an open sea exists?"

PART THREE

BEYOND THE NORTH WIND

I tried to forget my thirst by busying myself with bringing up some food and an empty vessel from the hold. Reaching over the side-rail, I filled the vessel with water for the purpose of laving my hands and face. To my astonishment, when the water came in contact with my lips, I could taste no salt. I was startled by the discovery. "Father!" I fairly gasped, "the water, the water; it is fresh!" "What, Olaf?" exclaimed my father, glancing hastily around. "Surely you are mistaken. There is no land. You are going mad." "But taste it!" I cried.

And thus, we made the discovery that the water was indeed fresh, absolutely so, without the least briny taste or even the suspicion of a salty flavor.

We forthwith filled our two-remaining water-casks, and my father declared it was a' heavenly dispensation of mercy from the gods Odin and Thor.

We were almost beside ourselves with joy, but hunger bade us end our enforced fast. Now that we had found fresh water in the open sea, what might we not expect in this strange latitude where ship had never before sailed, and the splash of an oar had never been heard? [1]

We had scarcely appeased our hunger when a breeze began filling the idle sails, and, glancing at the coin-pass, we found the northern point pressing hard against the glass.

In response to my surprise, my father said, "I have heard of this before; it is what they call the dipping of the needle."

We loosened the compass and turned it at right angles with the surface of the sea before its point would free itself from the glass and point according to unmolested attraction. It shifted uneasily, and seemed as unsteady as a drunken man, but finally pointed a course.

Before this we thought the wind was carrying us north by northwest, but, with the needle free, we discovered, if it could be relied upon, that we were sailing slightly north by northeast. Our course, however, was ever tending northward. [1]

The sea was serenely smooth, with hardly a choppy wave, and the wind brisk and exhilarating. The sun's rays, while striking us aslant, furnished tranquil warmth. And thus, time wore on day after day, and we found from the record in our log-book, we had been sailing eleven days since the storm in the open sea.

By strictest economy, our food was holding out fairly well, but beginning to run low. In the meantime, one of our casks of water had been exhausted, and my father said: "We will fill it again." But, to our dismay, we found the water was now as salt as in the region of the Lofoden Islands off the coast of Norway. This necessitated our being extremely careful of the remaining cask.

I found myself wanting to sleep much of the time; whether it was the effect of the exciting experience of sailing in unknown waters, or the relaxation from the awful excitement incident to our adventure in a storm at sea, or due to want of food, I could not say.

I frequently lay down on the bunker of our little sloop and looked far up into the blue dome of the sky; and, notwithstanding the sun was shining far away in the east, I always saw a single star overhead. For several days, when I looked for this star, it was always there directly above us.

It was now, according to our reckoning, about the first of August. The sun was high in the heavens and was so bright that I could no longer see the one lone star that attracted my attention a few days earlier.

One day about this time, my father startled me by calling my attention to a novel sight far in front of us, almost at the horizon. "It is a mock sun," exclaimed my father. "I have read of them; it is called a reflection or mirage. It will soon pass away."

But this dull-red, false sun, as we supposed it to be, did not pass away for several hours; and while we were unconscious of its emitting any rays of light, still there was no time thereafter when we could not sweep the horizon in front and locate the illumination of the so-called false sun, during a period of at least twelve hours out of every twenty-four.

"It could hardly be said to resemble the sun except in its circular shape."

Clouds and mists would at times almost, but never entirely, hide its location. Gradually it seemed to climb higher in the horizon of the uncertain purply sky as we advanced.

It could hardly be said to resemble the sun, except in its circular shape, and when not obscured by clouds or the ocean mists, it had a hazy-red, bronzed appearance, which would change to a white light like a luminous cloud, as if reflecting some greater light beyond.

We finally agreed in our discussion of this smoky furnace-colored sun, that, whatever the cause of the phenomenon, it was not a reflection of our sun, but a planet of some sort—a reality. [1]

One day soon after this, I felt exceedingly drowsy, and fell into a sound sleep. But it seemed that I was almost immediately aroused by my father's vigorous shaking of me by the shoulder and saying: "Olaf, awaken; there is land in sight!"

I sprang to my feet, and oh! joy unspeakable! There, far in the distance, yet directly in our path, were lands jutting boldly into the sea. The shoreline stretched far away to the right of us, as far as the eye could see, and all along the sandy beach were waves breaking into choppy foam, receding, then going forward again, ever chanting in monotonous thunder tones the song of the deep. The banks were covered with trees and vegetation.

I cannot express my feeling of exultation at this discovery. My father stood motionless, with his hand on the tiller, looking straight ahead, pouring out his heart in thankful prayer and thanksgiving to the gods Odin and Thor.

In the meantime, a net which we found in the stowage had been cast, and we caught a few fish that materially added to our dwindling stock of provisions.

The compass, which we had fastened back in its place, in fear of another storm, was still pointing due north, and moving on its pivot, just as it had at Stockholm. The dipping of the needle had ceased. What could this mean? Then, too, our many days of sailing had certainly carried us far past the North Pole. And yet the needle continued to point north. We were sorely perplexed, for surely our direction was now south. [1]

We sailed for three days along the shoreline, then came to the mouth of a fjord or river of immense size. It seemed more like a great bay, and into this we turned our fishing-craft, the direction being slightly northeast of south. By the assistance of a fretful wind that came to our aid about twelve hours out of every twenty-four, we continued to make our way inland, into what afterward proved to be a mighty river, and which we learned was called by the inhabitants Hiddekel.

We continued our journey for ten days thereafter and found we had fortunately attained a distance inland where ocean tides no longer affected the water, which had become fresh.

The discovery came none too soon, for our remaining cask of water was well-nigh exhausted. We lost no time in replenishing our casks and continued to sail farther up the river when the wind was favorable.

Along the banks great forests miles in extent could be seen stretching away on the shoreline. The trees were of enormous size. We landed after anchoring near a sandy beach, and waded ashore, and were rewarded by finding a quantity of nuts that were very palatable and satisfying to hunger, and a welcome change from the monotony of our stock of provisions.

It was about the first of September, over five months, we calculated, since our leave-taking from Stockholm. Suddenly we were frightened almost out of our wits by hearing in the far distance the singing of people. Very soon thereafter we discovered a huge ship gliding down the river directly toward us. Those aboard were singing in one mighty chorus that, echoing from bank to bank, sounded like a thousand voices, filling the whole universe with quivering melody. The accompaniment was played on stringed instruments not unlike our harps. It was a larger ship than any we bad ever seen and was differently constructed. [1]

At this particular time our sloop was becalmed, and not far from the shore. The bank of the river, covered with mammoth trees, rose up several hundred feet in beautiful fashion. We seemed to be on the edge of some primeval forest that doubtless stretched far inland.

The immense craft paused, and almost immediately a boat was lowered and six men of gigantic stature rowed to our little fishing-sloop. They spoke to us in a strange language. We knew from their manner, however, that they were not unfriendly. They talked a great deal among themselves, and one of them laughed immoderately, as though in finding us a queer discovery had been made. One of them spied our compass, and it seemed to interest them more than any other part of our sloop.

"They spoke to us in a strange language."

Finally, the leader motioned as if to ask whether we were willing to leave our craft to go on board their ship. "What say you, my son?" asked my father. "They cannot do any more than kill us."

"They seem to be kindly disposed," I replied, "although what terrible giants! They must be the select six of the kingdom's crack regiment. Just look at their great size."

"We may as well go willingly as be taken by force," said my father, smiling, "for they are certainly able to capture us." Thereupon he made known, by signs, that we were ready to accompany them.

Within a few minutes we were on board the ship, and half an hour later our little fishing-craft had been lifted bodily out of the water by a strange sort of hook and tackle and set on board as a curiosity.

There were several hundred people on board this, to us, mammoth ship, which we discovered was called "The Naz," meaning, as we afterward learned, "Pleasure," or to give a more proper interpretation, "Pleasure Excursion" ship.

If my father and I were curiously observed by the ship's occupants, this strange race of giants offered us an equal amount of wonderment.

There was not a single man aboard who would not have measured fully twelve feet in height. They all wore full beards, not particularly long, but seemingly short-cropped. They had mild and beautiful faces, exceedingly fair, with ruddy complexions. The hair and beard of some were black, others sandy, and still others yellow. The captain, as we designated the dignitary in command of the great vessel, was fully a head taller than any of his companions. The women averaged from ten to eleven feet in height. Their features were especially regular and refined, while their complexion was of a most delicate tint heightened by a healthful glow. [1]

Both men and women seemed to possess that particular ease of manner which we deem a sign of good breeding, and, notwithstanding their huge statures, there was nothing about them suggesting awkwardness. As I was a lad in only my nineteenth year, I was doubtless looked upon as a true Tom Thumb. My father's six feet three did not lift the top of his head above the waistline of these people.

Each one seemed to vie with the others in extending courtesies and showing kindness to us, but all laughed heartily, I remember, when they had to improvise chairs for my father and myself to sit at table. They were richly attired in a costume peculiar to themselves, and very attractive. The men were clothed in handsomely embroidered tunics of silk and satin and belted at the waist. They wore knee-breeches and stockings of a fine texture, while their feet were encased in sandals adorned with gold buckles. We early discovered that gold was one of the most common metals known, and that it was used extensively in decoration.

Strange as it may seem, neither my father nor myself felt the least bit of solicitude for our safety. "We have come into our own," my father said to me. "This is the fulfillment of the tradition told me by my father and my father's father, and still back for many generations of our race. This is, assuredly, the land beyond the North Wind."

We seemed to make such an impression on the party that we were given specially into the charge of one of the men, Jules Galdea, and his wife, for the purpose of being educated in their language; and we, on our part, were just as eager to learn as they were to instruct.

At the captain's command, the vessel was swung cleverly about, and began retracing its course up the river. The machinery, while noiseless, was very powerful.

The banks and trees on either side seemed to rush by. The ship's speed, at times, surpassed that of any railroad train on which I have ever ridden, even here in America. It was wonderful.

In the meantime, we had lost sight of the sun's rays, but we found a radiance "within" emanating from the dull-red sun which had already attracted our attention, now giving out a white light seemingly from a cloudbank far away in front of us. It dispensed a greater light, I should say, than two full moons on the clearest night.

In twelve hours, this cloud of whiteness would pass out of sight as if eclipsed, and the twelve hours following corresponded with our night. We early learned that these strange people were worshipers of this great cloud of night. It was "The Smoky God" of the "Inner World."

The ship was equipped with a mode of illumination which I now presume was electricity, but neither my father nor myself were sufficiently skilled in mechanics to understand whence came the power to operate the ship, or to maintain the soft beautiful lights that answered the same purpose of our present methods of lighting the streets of our cities, our houses and places of business.

It must be remembered, the time of which I write was the autumn of 1829, and we of the "outside" surface of the earth knew nothing then, so to speak, of electricity.

The electrically surcharged condition of the air was a constant vitalizer. I never felt better in my life than during the two years my father and I sojourned on the inside of the earth.

To resume my narrative of events: The ship on which we were sailing came to a stop two days after we had been taken on board. My father said as nearly as he could judge, we were directly under Stockholm or London.

The city we had reached was called "Jehu," signifying a seaport town. The houses were large and beautifully constructed, and quite uniform in appearance, yet without sameness. The principal occupation of the people appeared to be agriculture; the hillsides were covered with vineyards, while the valleys were devoted to the growing of grain.

I never saw such a display of gold. It was everywhere. The doorcasings were in laid and the tables were veneered with sheetings of gold. Domes of the public buildings were of gold. It was used most generously in the finishings of the great temples of music.

Vegetation grew in lavish exuberance, and fruit of all kinds possessed the most delicate flavor. Clusters of grapes four and five feet in length, each grape as large as an orange, and apples larger than a man's head typified the wonderful growth of all things on the "inside" of the earth.

The great redwood trees of California would be considered mere underbrush compared with the giant forest trees extending for miles and miles in all directions. In many directions along the foothills of the mountains vast herds of cattle were seen during the last day of our travel on the river.

We heard much of a city called "Eden," but were kept at "Jehu" for an entire year. By the end of that time, we had learned to speak fairly well the language of this strange race of people. Our instructors, Jules Galdea and his wife, exhibited a patience that was truly commendable.

One day an envoy from the Ruler at "Eden" came to see us, and for two whole days my father and myself were put through a series of surprising questions. They wished to know from whence we came, what sort of people dwelt "without," what God we worshiped, our religious beliefs, the mode of living in our strange land, and a thousand other things.

The compass which we had brought with us attracted especial attention. My father and I commented between ourselves on the fact that the compass still pointed north, although we now knew that we had sailed over the curve or edge of the earth's aperture and were far along southward on the "inside" surface of the earth's crust, which, according to my father's estimate and my own, is about three hundred miles in thickness from the "inside" to the "outside" surface. Relatively speaking, it is no thicker than an eggshell, so that there is almost as much surface on the "inside" as on the "outside" of the earth.

The great luminous cloud or ball of dull-red fire—fiery-red in the mornings and evenings, and during the day giving off a beautiful white light, "The Smoky God,"—is seemingly suspended in the center of the great vacuum "within" the earth, and held to its place by the immutable law of gravitation, or a repellant atmospheric force, as the case may be. I refer to the known power that draws or repels with equal force in all directions.

The base of this electrical cloud or central luminary, the seat of the gods, is dark and non-transparent, save for innumerable small openings, seemingly in the bottom of the great support or altar of the Deity, upon which "The Smoky God" rests; and, the lights shining through these many openings twinkle at night in all their splendor, and seem to be stars, as natural as the stars we saw shining when in our home at Stockholm, excepting that they appear larger. "The Smoky God," therefore, with each daily revolution of the earth, appears to come up in the east and go down in the west, the same as does our sun on the external surface. In reality, the people "within" believe that "The Smoky God" is the throne of their Jehovah and is stationary. The effect of night and day is, therefore, produced by the earth's daily rotation.

I have since discovered that the language of the people of the Inner World is much like the Sanskrit.

After we had given an account of ourselves to the emissaries from the central seat of government of the inner continent, and my father had, in his crude way, drawn maps, at their request, of the "outside" surface of the earth, showing the divisions of land and water, and giving the name of each of the continents, large islands and the oceans, we were taken overland to the city of "Eden," in a conveyance different from anything we have in Europe or America. This vehicle was doubtless some electrical contrivance. It was noiseless and ran on a single iron rail in perfect balance. The trip was made at a very high rate of speed. We were carried up hills and down dales, across valleys and again along the sides of steep mountains, without any apparent attempt having been made to level the earth as we do for railroad tracks. The car seats were huge yet comfortable affairs, and very high above the floor of the car. On the top of each car were high geared fly wheels lying on their sides, which were so automatically adjusted that, as the speed of the car increased, the high speed of these fly wheels geometrically increased.

Jules Galdea explained to us that these revolving fan-like wheels on top of the cars destroyed atmospheric pressure, or what is generally understood by the term gravitation, and with this force thus destroyed or rendered nugatory the car is as safe from falling to one side or the other from the single rail track as if it were in a vacuum; the fly wheels in their rapid revolutions destroying effectually the so-called power of gravitation, or the force of atmospheric pressure or whatever potent influence it may be that causes all unsupported things to fall downward to the earth's surface or to the nearest point of resistance.

The surprise of my father and myself was indescribable when, amid the regal magnificence of a spacious hall, we were finally brought before the Great High Priest, ruler over all the land. He was richly robed, and much taller than those about him, and could not have been less than fourteen or fifteen feet in height. The immense room in which we were received seemed finished in solid slabs of gold thickly studded with jewels of amazing brilliancy.

The city of "Eden" is located in what seems to be a beautiful valley, yet, in fact, it is on the loftiest mountain plateau of the Inner. Continent, several thousand feet higher than any portion of the surrounding country. It is the most beautiful place I have ever beheld in all my travels. In this elevated garden all manner of fruits, vines, shrubs, trees, and flowers grow in riotous profusion.

"We were brought before the Great High Priest."

In this garden four rivers have their source in a mighty artesian fountain. They divide and flow in four directions. This place is called by the inhabitants the "navel of the earth," or the beginning, "the cradle of the human race." The names of the rivers are the Euphrates, the Pison, the Gihon, and the Hiddekel.

The unexpected awaited us in this palace of beauty, in the finding of our little fishing-craft. It had been brought before the High Priest in perfect shape, just as it had been taken from the waters that day when it was loaded on board the ship by the people who discovered us on the river more than a year before.

We were given an audience of over two hours with this great dignitary, who seemed kindly disposed and considerate. He showed himself eagerly interested, asking us numerous questions, and invariably regarding things about which his emissaries had failed to inquire.

At the conclusion of the interview, he inquired our pleasure, asking us whether we wished to remain in his country or if we preferred to return to the "outer" world, providing it were possible to make a successful return trip, across the frozen belt barriers that encircle both the northern and southern openings of the earth.

My father replied: "It would please me and my son to visit your country and see your people, your colleges and palaces of music and art, your great fields, your wonderful forests of timber; and after we have had this pleasurable privilege, we should like to try to return to our home on the `outside' surface of the earth. This son is my only child, and my good wife will be weary awaiting our return."

"I fear you can never return," replied the Chief High Priest, "because the way is a most hazardous one. However, you shall visit the different countries with Jules Galdea as your escort and be accorded every courtesy and kindness. Whenever you are ready to attempt a return voyage, I assure you that your boat which is here on exhibition shall be put in the waters of the river Heddekel at its mouth, and we will bid you Jehovah-speed."

Thus terminated our only interview with the High Priest or Ruler of the continent.

Footnotes

30:1 *In vol. I, page* 196, *Nansen writes:* "It is a peculiar phenomenon, —this dead water. We had at present a better opportunity of studying it than we desired. It occurs where a surface layer of freshwater rests upon the salt water of the sea, and this fresh water is carried along with the ship gliding on the heavier sea beneath it as if on a fixed foundation. The difference between the two strata was in this case so great that while we had drinking water on the surface, the water we got from the bottom cock of the engine-room was far too salt to be used for the boiler."

31:1 *In volume II, pages* 18 *and* 19, *Nansen writes about the inclination of the needle. Speaking of Johnson, his aide:* "One day—it was November 24th—he came into supper a little after six o'clock, quite alarmed, and said: 'There has just been a singular inclination of the needle in twenty-four degrees. And remarkably enough, its northern extremity pointed to the east.'"

We again find in Peary's first voyage—page 67, *—the following:* "It had been observed that from the moment they had entered Lancaster Sound, the motion of the compass needle was very sluggish, and both this and its deviation increased as they progressed to the westward, and continued to do so in descending this inlet. Having reached latitude 73 degrees, they witnessed for the first time the curious phenomenon of the directive power of the needle becoming so weak as to be completely overcome by the attraction of the ship, so that the needle might now be said to point to the north pole of the ship."

33:1 *Nansen, on page 394, says: "Today another noteworthy thing happened, which was that about midday we saw the sun, or to be more correct, an image of the sun, for it was only a mirage. A peculiar impression was produced by the sight of that glowing fire lit just above the outermost edge of the ice. According to the enthusiastic descriptions given by many Arctic travelers of the first appearance of this god of life e after the long winter night, the impression ought to be one of jubilant excitement; but it was not so in my case. We had not expected to see it for some days yet, so that my feeling was rather one of pain, of disappointment, that we must have drifted farther south than we thought. So it was with pleasure I soon discovered that it could not be the sun itself. The mirage was at first a flattened-out, glowing red streak of fire on the horizon; later there were two streaks, the one above the other, with a dark space between; and from the maintop I could see four, or even five, such horizontal lines directly over one another, all of equal length, as if one could only imagine a square, dull-red sun, with horizontal dark streaks across it."*

34:1 *Peary's first voyage, pages 69 and 95 70, says: "On, reaching Sir Byam Martin's Island, the nearest to Melville Island, the latitude of the place of observation was 75 degrees-09´-23", and the longitude 103 degrees-44´-37"; the dip of the magnetic needle 88 degrees-25´-58" west in the longitude of 91 degrees-48´, where the last observations on the shore had been made, to 165 degrees-50´-09", east, at their present station, so that we had," says Peary, "in sailing over the space included between these two meridians, crossed immediately northward of the magnetic pole, and had undoubtedly passed over one of those spots upon the globe where the needle would have been found to vary 180 degrees, or in other words, where the North Pole would have pointed to the south."*

35:1 *Asiatic Mythology,—page 240, "Paradise Found"—from translation by Sayce, in a book called "Records of the Past," we were told of a "dwelling" which "the gods created for" the first human beings,—a dwelling in which they "became great" and "increased in numbers," and the location of which is described in words exactly corresponding to those of Iranian, Indian, Chinese, Eddaic and Aztecan literature; namely, "in the center of the earth."—Warren.*

37:1 *"According to all procurable data, that spot at the era of man's appearance upon the stage was in the now lost 'Miocene continent,' which then surrounded the Arctic Pole. That in that true, original Eden some of the early generations of men attained to a stature and longevity unequaled in any countries known to postdiluvian history is by no means scientifically incredible."—Wm. F. Warren, "Paradise Found," p. 284.*

43:1 *"And the Lord God planted a garden, and out of the ground made the Lord God to grow every tree that is pleasant to the sight and good for food."—The Book of Genesis.*

PART FOUR

IN THE UNDER WORLD

We learned that the males do not marry before they are from seventy-five to one hundred years old, and that the age at which women enter wedlock is only a little less, and that both men and women frequently live to be from six to eight hundred years old, and in some instances much older. [1]

During the following year we visited many villages and towns, prominent among them being the cities of Nigi, Delfi, Hectea, and my father was called upon no less than a half-dozen times to go over the maps which had been made from the rough sketches he had originally given of the divisions of land and water on the "outside" surface of the earth.

I remember hearing my father remark that the giant race of people in the land of "The Smoky God" had almost as accurate an idea of the geography of the "outside" surface of the earth as had the average college professor in Stockholm.

In our travels we came to a forest of gigantic trees, near the city of Delfi. Had the Bible said there were trees towering over three hundred feet in height, and more than thirty feet in diameter, growing in the Garden of Eden, the Ingersolls, the Tom Paines and Voltaires would doubtless have pronounced the statement a myth. Yet this is the description of the California *sequoia gigantea;* but these California giants pale into insignificance when compared with the forest Goliaths found in the "within" continent, where abound mighty trees from eight hundred to one thousand feet in height, and from one hundred to one hundred and twenty feet in diameter; countless in numbers and forming forests extending hundreds of miles back from the sea.

The people are exceedingly musical and learned to a remarkable degree in their arts and sciences, especially geometry and astronomy. Their cities are equipped with vast palaces of music, where not infrequently as many as twenty-five thousand lusty voices of this giant race swell forth in mighty choruses of the most sublime symphonies.

The children are not supposed to attend institutions of learning before they are twenty years old. Then their school life begins and continues for thirty years, ten of which are uniformly devoted by both sexes to the study of music.

Their principal vocations are architecture, agriculture, horticulture, the raising of vast herds of cattle, and the building of conveyances peculiar to that country, for travel on land and water. By some device which I cannot explain, they hold communion with one another between the most distant parts of their country, on air currents.

All buildings are erected with special regard to strength, durability, beauty and symmetry, and with a style of architecture vastly more attractive to the eye than any I have ever observed elsewhere.

About three-fourths of the "inner" surface of the earth is land and about one-fourth water. There are numerous rivers of tremendous size, some flowing in a northerly direction and others southerly. Some of these rivers are thirty miles in width, and it is out of these vast waterways, at the extreme northern and southern parts of the "inside" surface of the earth, in regions where low temperatures are experienced, that freshwater icebergs are formed. They are then pushed out to sea like huge tongues of ice, by the abnormal freshets of turbulent waters that, twice every year, sweep everything before them.

We saw innumerable specimens of birdlife no larger than those encountered in the forests of Europe or America. It is well known that during the last few years whole species of birds have quit the earth. A writer in a recent article on this subject says: [1]

Is it not possible that these disappearing bird species quit their habitation without, and find an asylum in the "within world"?

Whether inland among the mountains, or along the seashore, we found bird life prolific. When they spread their great wings some of the birds appeared to measure thirty feet from tip to tip. They are of great variety and many colors. We were permitted to climb up on the edge of a rock and examine a nest of eggs. There were five in the nest, each of which was at least two feet in length and fifteen inches in diameter.

After we had been in the city of Hectea about a week, Professor Galdea took us to an inlet, where we saw thousands of tortoises along the sandy shore. I hesitate to state the size of these great creatures. They were from twenty-five to thirty feet in length, from fifteen to twenty feet in width and fully seven feet in height. When one of them projected its head, it had the appearance of some hideous sea monster.

The strange conditions "within" are favorable not only for vast meadows of luxuriant grasses, forests of giant trees, and all manner of vegetable life, but wonderful animal life as well.

"There must have been five hundred of these thunder-throated monsters."

One day we saw a great herd of elephants. There must have been five hundred of these thunder-throated monsters, with their restlessly waving trunks. They were tearing huge boughs from the trees and trampling smaller growth into dust like so much hazel-brush. They would average over 100 feet in length and from 75 to 85 in height.

It seemed, as I gazed upon this wonderful herd of giant elephants, that I was again living in the public library at Stockholm, where

I had spent much time studying the wonders of the Miocene age. I was filled with mute astonishment, and my father was speechless with awe. He held my arm with a protecting grip, as if fearful harm would overtake us. We were two atoms in this great forest, and, fortunately, unobserved by this vast herd of elephants as they drifted on and away, following a leader as does a herd of sheep. They browsed from growing herbage which they encountered as they traveled, and now and again shook the firmament with their deep bellowing. 1

There is a hazy mist that goes up from the land each evening, and it invariably rains once every twenty-four hours. This great moisture and the invigorating electrical light and warmth account perhaps for the luxuriant vegetation, while the highly charged electrical air and the evenness of climatic conditions may have much to do with the giant growth and longevity of all animal life.

In places the level valleys stretched away for many miles in every direction. "The Smoky God," in its clear white light, looked calmly down. There was an intoxication in the electrically surcharged air that fanned the cheek as softly as a vanishing whisper. Nature chanted a lullaby in the faint murmur of winds whose breath was sweet with the fragrance of bud and blossom.

After having spent considerably more than a year in visiting several of the many cities of the "within" world and a great deal of intervening country, and more than two years had passed from the time we had been picked up by the great excursion ship on the river, we decided to cast our fortunes once more upon the sea, and endeavor to regain the "outside" surface of the earth.

We made known our wishes, and they were reluctantly but promptly followed. Our hosts gave my father, at his request, various maps showing the entire "inside" surface of the earth, its cities, oceans, seas, rivers, gulfs and bays. They also generously offered to give us all the bags of gold nuggets—some of them as large as a goose's egg—that we were willing to attempt to take with us in our little fishing-boat.

In due time we returned to Jehu, at which place we spent one month in fixing up and overhauling our little fishing sloop. After all was in readiness, the same ship "Naz" that originally discovered us, took us on board and sailed to the mouth of the river Hiddekel.

After our giant brothers had launched our little craft for us, they were most cordially regretful at parting, and evinced much solicitude for our safety. My father swore by the Gods Odin and Thor that he would surely return again within a year or two and pay them another visit. And thus, we bade them adieu. We made ready and hoisted our sail, but there was little breeze. We were becalmed within an hour after our giant friends had left us and started on their return trip.

The winds were constantly blowing south, that is, they were blowing from the northern opening of the earth toward that which we knew to be south, but which, according to our compass's pointing finger, was directly north.

For three days we tried to sail, and to beat against the wind, but to no avail. Whereupon my father said: "My son, to return by the same route as we came in is impossible at this time of year. I wonder why we did not think of this before. We have been here almost two and a half years; therefore, this is the season when the sun is beginning to shine in at the southern opening of the earth. The long cold night is on in the Spitzbergen country."

"What shall we do?" I inquired.

"There is only one thing we can do," my father replied, "and that is to go south." Accordingly, he turned the craft about, gave it full reef, and started by the compass north but, in fact, directly south. The wind was strong, and we seemed to have struck a current that was running with remarkable swiftness in the same direction.

In just forty days we arrived at Delfi, a city we had visited in company with our guides Jules Galdea and his wife, near the mouth of the Gihon river. Here we stopped for two days and were most hospitably entertained by the same people who had welcomed us on our former visit. We laid in some additional provisions and again set sail, following the needle due north.

On our outward trip we came through a narrow channel which appeared to be a separating body of water between two considerable bodies of land. There was a beautiful beach to our right, and we decided to reconnoiter. Casting anchor, we waded ashore to rest up for a day before continuing the outward hazardous undertaking. We built a fire and threw on some sticks of dry driftwood. While my father was walking along the shore, I prepared a tempting repast from supplies we had provided.

There was a mild, luminous light which my father said resulted from the sun shining in from the south aperture of the earth. That night we slept soundly, and awakened the next morning as refreshed as if we had been in our own beds at Stockholm.

After breakfast we started out on an inland tour of discovery but had not gone far when we sighted some birds which we recognized at once as belonging to the penguin family. They are flightless birds, but excellent swimmers and tremendous in size, with white breast, short wings, black head, and long peaked bills. They stand fully nine feet high. They looked at us with little surprise, and presently waddled, rather than walked, toward the water, and swam away in a northerly direction. 1

The events that occurred during the following hundred or more days beggar description. We were on an open and iceless sea. The month we reckoned to be November or December, and we knew the so-called South Pole was turned toward the sun. Therefore, when passing out and away from the internal electrical light of "The Smoky God" and its genial warmth, we would be met by the light and warmth of the sun, shining in through the south opening of the earth. We were not mistaken. 2

There were times when our little craft, driven by wind that was continuous and persistent, shot through the waters like an arrow. Indeed, had we encountered a hidden rock or obstacle our little vessel would have been crushed into kindling-wood.

At last, we were conscious that the atmosphere was growing decidedly colder and a few days later, icebergs were sighted far to the left. My father argued, and correctly, that the winds which filled our sails came from the warm climate "within." The time of the year was certainly most auspicious for us to make our dash for the "outside" world and attempt to scud our fishing sloop through open channels of the frozen zone which surrounds the polar regions.

We were soon amid the icepacks, and how our little craft got through the narrow channels and escaped being crushed I know not. The compass behaved in the same drunken and unreliable fashion in passing over the southern curve or edge of the earth's shell as it had done on our inbound trip at the northern entrance. It gyrated, dipped and seemed like a thing possessed. 3

One day as I was lazily looking over the sloop's side into the clear waters, my father shouted: "Breakers ahead!" Looking up, I saw through a lifting mist a white object that towered several hundred feet high, completely shutting off our advance. We lowered sail immediately, and none too soon. In a moment we found ourselves wedged between two monstrous icebergs. Each was crowding and grinding against its fellow mountain of ice. They were like two gods of war contending for supremacy. We were greatly alarmed. Indeed, we were between the lines of a battle royal; the sonorous thunder of the grinding ice was like the continued volleys of artillery. Blocks of ice larger than a house were frequently lifted up a hundred feet by the mighty force of lateral pressure; they would shudder and rock to and fro for a few seconds, then come crashing down with a deafening roar, and disappear in the foaming waters. Thus, for more than two hours, the contest of the icy giants continued.

"My father shouted: 'Breakers ahead!'

It seemed as if the end had come. The ice pressure was terrific, and while we were not caught in the dangerous part of the jam, and were safe for the time being, yet the heaving and rending of tons of ice as it fell splashing here and there into the watery depths filled us with shaking fear.

Finally, to our great joy, the grinding of the ice ceased, and within a few hours the great mass slowly divided, and, as if an act of Providence had been performed, right before us lay an open channel. Should we venture with our little craft into this opening? If the pressure came on again, our little sloop as well as ourselves would be crushed into nothingness. We decided to take the chance, and, accordingly, hoisted our sail to a favoring breeze, and soon started out like a racehorse, running the gauntlet of this unknown narrow channel of open water.

Footnotes

48:1 *Josephus says: "God prolonged the life of the patriarchs that preceded the deluge, both on account of their virtues and to give them the opportunity of perfecting the sciences of geometry and astronomy, which they had discovered; which they could not have done if they had not lived 600 years, because it is only after the lapse of 600 years that the great year is accomplished."—Flammarion, Astronomical Myths, Paris p. 26.*

50:1 *"Almost every year sees the final extinction of one or more bird species. Out of fourteen varieties of birds found a century since on a single island—the West Indian island of St. Thomas—eight have now to be numbered among the missing."*

52:1 *"Moreover, there were a great number of elephants in the island: and there was provision for animals of every kind. Also whatever fragrant things there are in the earth, whether roots or* herbage, or woods, or distilling drops of flowers or fruits, grew and thrived in that land."—The Cratyluo of Plato.

55:1 "The nights are never so dark at the Poles as in other regions, for the moon and stars seem to possess twice as much light and effulgence. In addition, there is a continuous light, the varied shades and play of which are amongst the strangest phenomena of nature."—Rambrosson's Astronomy.

55:2 "The fact that gives the phenomenon of the polar aurora its greatest importance is that the earth becomes self-luminous; that, besides the light which as a planet is received from the central body, it shows a capability of sustaining a luminous process proper to itself."—Humboldt.

55:3 Captain Sabine, on page 105 in "Voyages in the Arctic Regions," says: "The geographical determination of the direction and intensity of the magnetic forces at different points of the earth's surface has been regarded as an object worthy of especial research. To examine in different parts of the globe, the declination, inclination and intensity of the magnetic force, and their periodical and secular variations, and mutual relations and dependencies could be duly investigated only in fixed magnetical observatories."

PART FIVE

AMONG THE ICE PACKS

For the next forty-five days our time was employed in dodging icebergs and hunting channels; indeed, had we not been favored with a strong south wind and a small boat, I doubt if this story could have ever been given to the world.

At last, there came a morning when my father said: "My son, I think we are to see home. We are almost through the ice. See! the open water lies before us."

However, there were a few icebergs that had floated far northward into the open water still ahead of us on either side, stretching away for many miles. Directly in front of us, and by the compass, which had now righted itself, due north, there was an open sea.

"What a wonderful story we have to tell the people of Stockholm," continued my father, while a look of pardonable elation lighted up his honest face. "And think of the gold nuggets stowed away in the hold!"

I spoke kind words of praise to my father, not alone for his fortitude and endurance, but also for his courageous daring as a discoverer, and for having made the voyage that now promised a successful end. I was grateful, too, that he had gathered the wealth of gold we were carrying home.

While congratulating ourselves on the goodly supply of provisions and water we still had on hand, and on the dangers we had escaped, we were startled by hearing a most terrific explosion, caused by the tearing apart of a huge mountain of ice. It was a deafening roar like the firing of a thousand cannon. We were sailing at the time with great speed and happened to be near a monstrous iceberg which to all appearances was as immovable as a rockbound island. It seemed, however, that the iceberg had split and was breaking apart, whereupon the balance of the monster along which we were sailing was destroyed, and it began dipping from us. My father quickly anticipated the danger before I realized its awful possibilities. The iceberg extended down into the water many hundreds of feet, and, as it tipped over, the portion coming up out of the water caught our fishing-craft like a lever on a fulcrum and threw it into the air as if it had been a football.

Our boat fell back on the iceberg, that by this time had changed the side next to us for the top. My father was still in the boat, having become entangled in the rigging, while I was thrown some twenty feet away.

I quickly scrambled to my feet and shouted to my father, who answered: "All is well." Just then a realization dawned upon me. Horror upon horror! The blood froze in my veins.

The iceberg was still in motion, and its great weight and force in toppling over would cause it to submerge temporarily. I fully realized what a sucking maelstrom it would produce amid the worlds of water on every side. They would rush into the depression in all their fury, like white-fanged wolves eager for human prey.

In this supreme moment of mental anguish, I remember glancing at our boat, which was lying on its side, and wondering if it could possibly right itself, and if my father could escape. Was this the end of our struggles and adventures? Was this death? All these questions flashed through my mind in the fraction of a second, and a moment later I was engaged in a life and death struggle. The ponderous monolith of ice sank below the surface, and the frigid waters gurgled around me in frenzied anger. I was in a saucer, with the waters pouring in on every side. A moment more and I lost consciousness.

When I partially recovered my senses and roused from the swoon of a half-drowned man, I found myself wet, stiff, and almost frozen, lying on the iceberg. But there was no sign of my father or of our little fishing sloop. The monster berg had recovered itself, and, with its new balance, lifted its head perhaps fifty feet above the waves. The top of this island of ice was a plateau perhaps half an acre in extent.

I loved my father well and was grief-stricken at the awfulness of his death. I railed at fate, that I, too, had not been permitted to sleep with him in the depths of the ocean. Finally, I climbed to my feet and looked about me. The purple-domed sky above, the shoreless green ocean beneath, and only an occasional iceberg discernible! My heart sank in hopeless despair. I cautiously picked my way across the berg toward the other side, hoping that our fishing craft had righted itself.

Dared I think it possible that my father still lived? It was but a ray of hope that flamed up in my heart. But the anticipation warmed my blood in my veins and started it rushing like some rare stimulant through every fiber of my body.

I crept close to the precipitous side of the iceberg, and peered far down, hoping, still hoping. Then I made a circle of the berg, scanning every foot of the way, and thus I kept going around and around. One part of my brain was certainly becoming maniacal, while the other part, I believe, and due to this day, was perfectly rational.

I was conscious of having made the circuit a dozen times, and while one part of my intelligence knew, in all reason, there was not a vestige of hope, yet some strange fascinating aberration bewitched and compelled me still to beguile myself with expectation. The other part of my brain seemed to tell me that while there was no possibility of my father being alive, yet, if I quit making the circuitous pilgrimage, if I paused for a single moment, it would be acknowledgment of defeat, and, should I do this, I felt that I should go mad. Thus, hour after hour I walked around and around, afraid to stop and rest, yet physically powerless to continue much longer. Oh! horror of horrors! to be cast away in this wide expanse of waters without food or drink, and only a treacherous iceberg for an abiding place. My heart sank within me, and all semblance of hope was fading into black despair.

Then the hand of the Deliverer was extended, and the death-like stillness of a solitude rapidly becoming unbearable was suddenly broken by the firing of a signal-gun. I looked up in startled amazement, when, I saw, less than a half-mile away, a whaling-vessel bearing down toward me with her sail full set.

Evidently my continued activity on the iceberg had attracted their attention. On drawing near, they put out a boat, and, descending cautiously to the water's edge, I was rescued, and a little later lifted on board the whaling-ship.

I found it was a Scotch whaler, "The Arlington." She had cleared from Dundee in September, and started immediately for the Antarctic, in search of whales. The captain, Angus MacPherson, seemed kindly disposed, but in matters of discipline, as I soon learned, possessed of an iron will. When I attempted to tell him that I had come from the "inside" of the earth, the captain and mate looked at each other, shook their heads, and insisted on my being put in a bunk under strict surveillance of the ship's physician.

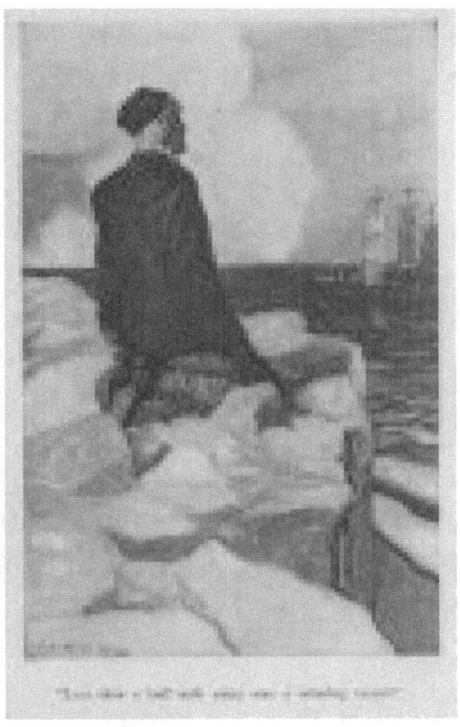

"Less than a half mile away was a whaling vessel."

I was very weak for want of food and had not slept for many hours. However, after a few days' rest, I got up one morning and dressed myself without asking permission of the physician or anyone else and told them that I was as sane as anyone.

The captain sent for me and again questioned me concerning where I had come from, and how I came to be alone on an iceberg in the far-off Antarctic Ocean. I replied that I had just come from the "inside" of the earth and proceeded to tell him how my father and myself had gone in by way of Spitzbergen and come out by way of the South Pole country, whereupon I was put in irons. I afterward heard the captain tell the mate that I was as crazy as a March hare, and that I must remain in confinement until I was rational enough to give a truthful account of myself.

Finally, after much pleading and many promises, I was released from irons. I then and there decided to invent some story that would satisfy the captain, and never again refer to my trip to the land of "The Smoky God," at least until I was safe among friends.

Within a fortnight I was permitted to go about and take my place as one of the seamen. A little later the captain asked me for an explanation. I told him that my experience had been so horrible that I was fearful of my memory and begged him to permit me to leave the question unanswered until sometime in the future. "I think you are recovering considerably," he said, "but you are not sane yet by a good deal." "Permit me to do such work as you may assign," I replied, "and if it does not compensate you sufficiently, I will pay you immediately after I reach Stockholm—to the last penny." Thus, the matter rested.

"Whereupon I was put in irons."

On finally reaching Stockholm, as I have already related, I found that my good mother had gone to her reward more than a year before. I have also told how, later, the treachery of a relative landed me in a mad-house, where I remained for twenty-eight years—seemingly unending years—and, still later, after my release, how I returned to the life of a fisherman, following it sedulously for twenty-seven years, then how I came to America, and finally to Los Angeles, California. But all this can be of little interest to the reader. Indeed, it seems to me the climax of my wonderful travels and strange adventures was reached when the Scotch sailing-vessel took me from an iceberg on the Antarctic Ocean.

PART SIX

CONCLUSION

In concluding this history of my adventures, I wish to state that I firmly believe science is yet in its infancy concerning the cosmology of the earth. There is so much that is unaccounted for by the world's accepted knowledge of today and will ever remain so until the land of "The Smoky God" is known and recognized by our geographers.

It is the land from whence came the great logs of cedar that have been found by explorers in open waters far over the northern edge of the earth's crust, and also the bodies of mammoths whose bones are found in vast beds on the Siberian coast.

Northern explorers have done much. Sir John Franklin, De Haven Grinnell, Sir John Murray, Kane, Melville, Hall, Nansen, Schwatka, Greely, Peary, Ross, Gerlache, Bernacchi, Andree, Amsden, Amundson and others have all been striving to storm the frozen citadel of mystery.

I firmly believe that Andree and his two brave companions, Strindberg and Fraenckell, who sailed away in the balloon "Oreon" from the northwest coast of Spitzbergen on that Sunday afternoon of July 11, 1897, are now in the "within" world, and doubtless are being entertained, as my father and myself were entertained by the kind-hearted giant race inhabiting the inner Atlantic Continent.

Having, in my humble way, devoted years to these problems, I am well acquainted with the accepted definitions of gravity, as well as the cause of the magnetic needle's attraction, and I am prepared to say that it is my firm belief that the magnetic needle is influenced solely by electric currents which completely envelop the earth like a garment, and that these electric currents in an endless circuit pass out of the southern end of the earth's cylindrical opening, diffusing and spreading themselves over all the "outside" surface, and rushing madly on in their course toward the North Pole. And while these currents seemingly dash off into space at the earth's curve or edge, yet they drop again to the "inside" surface and continue their way southward along the inside of the earth's crust, toward the opening of the so-called South Pole. [1]

As to gravity, no one knows what it is, because it has not been determined whether it is atmospheric pressure that causes the apple to fall, or whether, 150 miles below the surface of the earth, supposedly one halfway through the earth's crust, there exists some powerful loadstone attraction that draws it. Therefore, whether the apple, when it leaves the limb of the tree, is drawn or impelled downward to the nearest point of resistance, is unknown to the students of physics.

Sir James Ross claimed to have discovered the magnetic pole at about seventy-four degrees latitude. This is wrong—the magnetic pole is exactly one-half the distance through the earth's crust. Thus, if the earth's crust is three hundred miles in thickness, which is the distance I estimate it to be, then the magnetic pole is undoubtedly one hundred and fifty miles below the surface of the earth, it matters not where the test is made. And at this particular point one hundred and fifty miles below the surface, gravity ceases, becomes neutralized; and when we pass beyond that point on toward the "inside" surface of the earth, a reverse attraction geometrically increases in power, until the other one hundred and fifty miles of distance is traversed, which would bring us out on the "inside" of the earth.

Thus, if a hole were bored down through the earth's crust at London, Paris, New York, Chicago, or Los-Angeles, a distance of three hundred miles, it would connect the two surfaces.

While the inertia and momentum of a weight dropped in from the "outside" surface would carry it far past the magnetic center, yet, before reaching the "inside" surface of the earth it would gradually diminish in speed, after passing the halfway point, finally pause and immediately fall back toward the "outside" surface, and continue thus to oscillate, like the swinging of a pendulum with the power removed, until it would finally rest at the magnetic center, or at that particular point exactly one-half the distance between the "outside" surface and the "inside" surface of the earth.

The gyration of the earth in its daily act of whirling around in its spiral rotation—at a rate greater than one thousand miles every hour, or about seventeen miles per second—makes of it a vast electro-generating body, a huge machine, a mighty prototype of the puny-man-made dynamo, which, at best, is but a feeble imitation of nature's original.

The valleys of this inner Atlantis Continent, bordering the upper waters of the farthest north are in season covered with the most magnificent and luxuriant flowers. Not hundreds and thousands, but millions, of acres, from which the pollen or blossoms are carried far away in almost every direction by the earth's spiral gyrations and the agitation of the wind resulting therefrom, and it is these blossoms or pollen from the vast floral meadows "within" that produce the colored snows of the Arctic regions that have so mystified the northern explorers. [1]

Beyond question, this new land "within" is the home, the cradle, of the human race, and viewed from the standpoint of the discoveries made by us, must of necessity have a most important bearing on all physical, paleontological, archæological, philological and mythological theories of antiquity.

The same idea of going back to the land of mystery—to the very beginning—to the origin of man—is found in Egyptian traditions of the earlier terrestrial regions of the gods, heroes and men, from the historical fragments of Manetho, fully verified by the historical records taken from the more recent excavations of Pompeii as well as the traditions of the North American Indians.

It is now one hour past midnight—the new year of 1908 is here, and this is the third day thereof, and having at last finished the record of my strange travels and adventures I wish given to the world, I am ready, and even longing, for the peaceful rest which I am sure will follow life's trials and vicissitudes. I am old in years, and ripe both with adventures and sorrows, yet rich with the few friends I have cemented to me in my struggles to lead a just and upright life. Like a story that is well-nigh told, my life is ebbing away. The presentiment is strong within me that I shall not live to see the rising of another sun. Thus, do I conclude my message.

Olaf Jansen.

Footnotes

67:1 *"Mr. Lemstrom concluded that an electric discharge which could only be seen by means of the spectroscope was taking place on the surface of the ground all around him, and that from a distance it would appear as a faint display of Aurora, the phenomena of pale and flaming light which is some times seen on the top of the Spitzbergen Mountains."—The Arctic Manual, page* 739.

68:1 *Kane, vol. I, page* 44, *says: "We passed the 'crimson cliffs' of Sir John Ross in the forenoon of August 5th. The patches of red snow from which they derive their name could be seen clearly at the distance of ten miles from the coast."*

La Chambre, in an account of Andree's balloon expedition, on page 144, *says: "On the isle of Amsterdam the snow is tinted with red for a considerable distance, and the savants are collecting it to examine it microscopically. It presents, in fact, certain peculiarities; it is thought that it contains very small plants. Scoreby, the famous whaler, had already remarked this."*

PART SEVEN

AUTHOR'S AFTERWORD

I found much difficulty in deciphering and editing the manuscripts of Olaf Jansen. However, I have taken the liberty of reconstructing only a very few expressions, and in doing this have in no way changed the spirit or meaning. Otherwise, the original text has neither been added to nor taken from.

It is impossible for me to express my opinion as to the value or reliability of the wonderful statements made by Olaf Jansen. The description here given of the strange lands and people visited by him, location of cities, the names and directions of rivers, and other information herein combined, conform in every way to the rough drawings given into my custody by this ancient Norseman, which drawings together with the manuscript it is my intention at some later date to give to the Smithsonian Institution, to preserve for the benefit of those interested in the mysteries of the "Farthest North "—the frozen circle of silence. It is certain there are many things in Vedic literature, in "Josephus," the "Odyssey," the "Iliad," Terrien de Lacouperie's "Early History of Chinese Civilization," Flammarion's "Astronomical Myths," "Lenormant's "Beginnings of History," Hesiod's "Theogony,"

Sir John de Maundeville's writings, and Sayce's "Records of the Past," that, to say the least, are strangely in harmony with the seemingly incredible text found in the yellow manuscript of the old Norseman, Olaf Jansen, and now for the first time given to the world.

BOOK 2:

Symzonia
Voyage of Discovery

Captain Adam Seaborn

[1820]

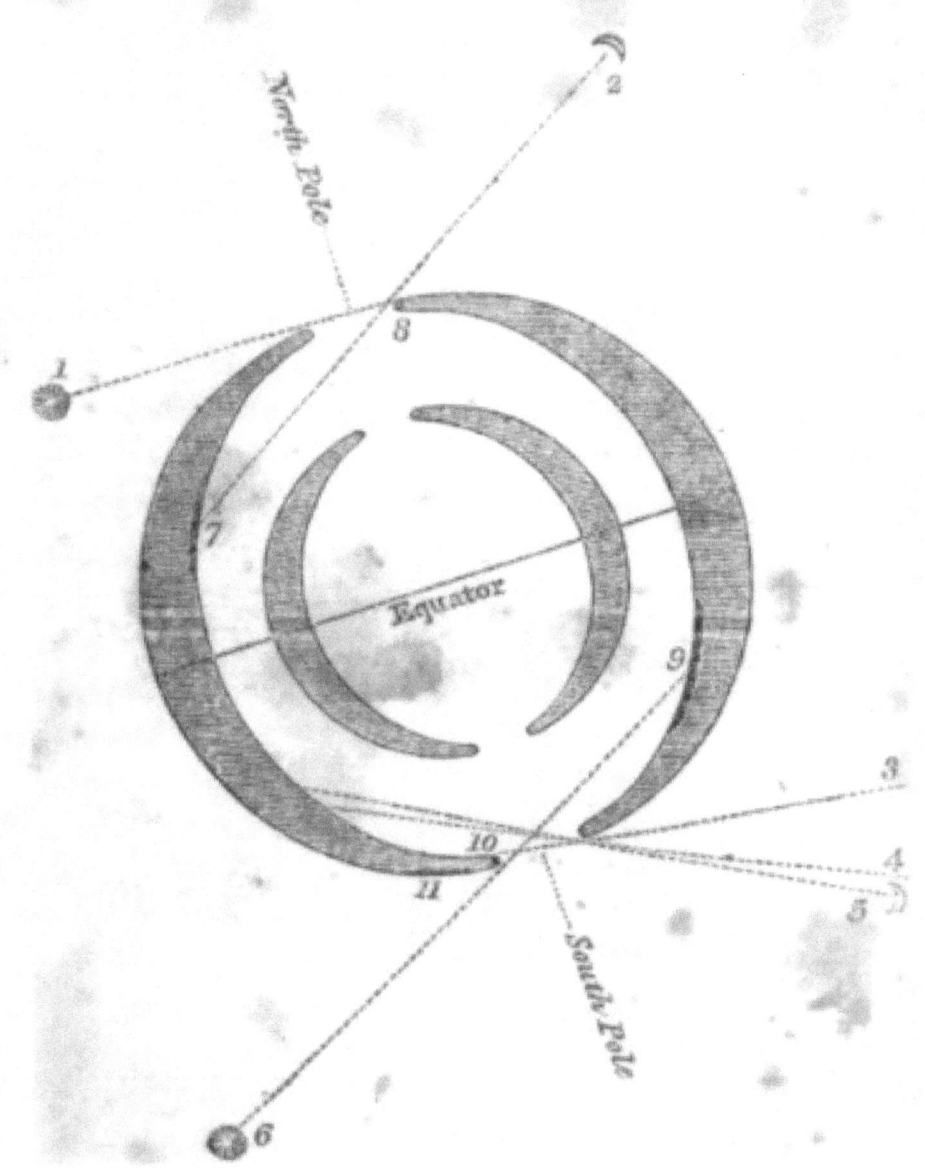

This is among the first and rarest of the Hollow Earth books. In the early 19th century, John Cleves Symmes proposed that the earth was actually a set of five nested spheres, each with polar openings. He unsuccessfully lobbied congress to fund an expedition to the poles to investigate this theory. This novel, based on Symmes' theories, was published in 1820. Some believe that Symmes was the author, but this has not been established beyond a shadow of doubt.

Probably the first US utopian science fiction novel, Symzonia is also a classic sea yarn. The narrator, Mr. Seaborn, initiates just such an expedition as Symmes dreamed of. With a special high-tech paddle boat, built with no metal to avoid magnetic effects, Seaborn sets out on a voyage into uncharted southern waters. Conflict is provided by a mutinous Mr. Slim, who proves to be Seaborn's nemesis. Eventually the ship penetrates into the Antarctic hole, and voyages deep into the inner surface of the earth, which Seaborn claims for the US and calls 'Symzonia'. They encounter a utopian society, run by a benevolent aristocracy, which has long since banished all of their misfits to a distant colony, and possibly the outside Earth. The Symzonians have advanced airships, boats with jet propulsion, flame-proof cloth woven from spiderwebs, and a mysterious weapon of mass destruction. Eventually, they expel Seaborn and his crew back to the exterior world. Due to a series of accidents, all material evidence of Symzonia is lost on the return voyage, and Seaborn is swindled out of his profits.

Edmond Halley, the famed astronomer, was the first to propose that the Earth contained a set of nested hollow spheres. Symme's theory, and this book in particular, was likely one of Poe's primary influences, particularly seen in his Narrative of Arthur Gordon Pym. The rumor of a hidden advanced civilization inside the earth has long been elaborated in fictional and occult books from Bulwer-Lytton to the the Shaver Mystery.

PRODUCTION NOTES: Published in 1820 and only reprinted once in 1965, this book is almost impossible to obtain. I searched for years for a copy of this in vain. I was finally able to obtain a copy of the 1965 facsimile for over $250, and used that as the basis of this etext.--J.B. Hare, December 8, 2008.

TABLE OF CONTENTS

CHAPTER I.

The Author's reasons for undertaking a voyage of discovery.—He builds a vessel for his purpose upon a new plan.—His departure from the United States.

Page 83

CHAPTER II.

The Author arrives at the Falkland Islands—Describes West Point Island, and States harbour—Visits the city of the Gentoo Penguins on the Grand Jason—Gives some account of the polity and habits of those civilized amphibia—Sails for South Georgia.

Page 88

CHAPTER III.

The Author passes South Georgia, and proceeds in search of Sandwich land—States to his officers and men his reasons for believing in the existence of great bodies of land within the antarctic circle, and for the opinion that the polar region is subject to great heat in summer. —Crew mutiny at the instigation of Mr. Slim, third mate.—Happy discovery of a southern continent, which, at the unanimous and earnest solicitation of his officers and men, he names Seaborn's Land.

Page 97

CHAPTER IV.

The Author in great peril, from the vast rise and fall of the tide in the polar sea—Brief account of his observations at Seaborn's Land—He takes formal possession of the country, in the manner usual in such eases, in the name and on behalf of the United States—Leaves a sealing party on one of the islands near the coast, and proceeds to the south, to extend his discoveries.

Page 109

CHAPTER V.

The Author discovers the south extremity of Seaborn's Land, which he names Cape World send. —The compass becomes useless.—He states the manner in which he obviated the difficulty occasioned thereby.—He enters the internal world: describes the phenomena which occur.—Discovers Token Island.—Occurrences at that Island.

Page 120

CHAPTER VI.

The Author departs from Token Island, in search of an internal continent. —Wind, weather, and other phenomena of the internal seas.—Great alarm of the crew.—Discovery of an inhabited country

Page 128

CHAPTER VII.

Description of the first view of the coast. —The Author names the discovered country Symzonia.—Enters the harbour.—His first interview with the Symzonian.—Sketch of their appearance.—He commences the study of the Symzonian language.—Wonderful powers of mind displayed by the natives.—Account of an aerial vessel.

Page 134

CHAPTER VIII.

The Author leaves the ship to visit the seat of government. — Description of the country.—Account of the polity of the Symzonians, as stated by his conductor.—Comparison of the industry, its objects and ends in the two worlds, and of the necessities and habits of the internals and externals.—Expulsion of the unworthy from Symzonia, to a place of exile near the north pole.—External world supposed to have been peopled by the outcasts.

Page 145

CHAPTER IX.

The Author arrives at the seat of government. —Description of the Auditory. —Symzonian manner of assembling for devotion and public business.—Etiquette of the Symzonian Court.—He is admitted to an audience by the Best Man.—Account of the interview, and of his unfortunate efforts to exalt the character of the externals, by describing some of their splendid follies.

Page 156

CHAPTER X.

Containing some account of the strange rationality of the Symzonians. —Their simplicity of dress. —Manner of making cloth. —Circulating medium. —Taxes.

Page 168

CHAPTER XI.

Containing some account of the Symzonian engine of defence.—Story of a very ancient war with an internal nation called Belzubians, which caused the invention of this engine.—Opposition of the Good men to its being used.—Fultria the inventor's speech in defence of it.—Deliberations of the Council.—Termination of the war.—Sentiments of the people on the subject.

Page 173

CHAPTER XII.

Wonderful faculties of the Symzonians. —Translation of my books into their language.—Proposition of a Wise man to make slaves of the Author and his people.—The Author's remonstrance.—The Wise man disgraced.

Page 180

CHAPTER XIII.

Recreations of the Symzonians. —Wonderful provision of nature for supplying the internal world with light.—Character and employments of the women of Symzonia.

Page 183

CHAPTER XIV.

The Author examines the records of the Assembly. —Grounds of proposal for admittance to the order of Worthies.—Shell fish of Symzonia.—Great quantities of Pearls, and the use to which they are applied.

Page 187

CHAPTER XV.

The Author is ordered to depart from Symzonia. —The Best Man's reasons for sending him away. — His ineffectual efforts to obtain a place of rendezvous for purposes of trade.

Page 191

CHAPTER XVI.

The Author returns to the Explorer—Holds a council of officers—Determines to return to Seaborn's Land—Takes leave of Surui—Sails from Symzonia—Touches at Token Island—Arrives at Boneto's station.

Page 197

CHAPTER XVII.

The Author loads the Explorer with seal skins, and sails from Seaborn's Land—Discovers Albicore's Islands. —Transactions at those islands. —He determines to conceal his discoveries from the world.—His reasons for this determination, and measures to effect it.—Sails for Canton.

Page 204

CHAPTER XVIII.

The Author arrives at Canton. —Transactions in China.—Sails for the United States.—Loss of manuscripts.—Difficulties with Mr. Slim.

Page 208

CHAPTER XIX.

Hurricane off the Isle of France. —Its consequences.—Death of Mr. Slim.

Page 212

CHAPTER XX.

The Author arrives in the United States—Consigns his cargo to Mr. Slippery—Is reduced to poverty by the failure of Mr. Slippery. —His great distress. —Inducement to publish this brief account of his discoveries. —Conclusion.

Page 215

A VOYAGE.

CHAPTER I.

The Author's reasons for undertaking a voyage of discovery. — He builds a vessel for his purpose upon a new plan.—His departure from the United States.

In the year 1817, I projected a voyage of discovery, in the hope of finding a passage to a new and untried world. I flattered myself that I should open the way to new fields for the enterprise of my fellow-citizens, supply new sources of wealth, fresh food for curiosity, and additional means of enjoyment; objects of vast importance, since the resources of the known world have been exhausted by research, its wealth monopolized, its wonders of curiosity explored, it's everything investigated and understood!

The state of the civilized world, and the growing evidence of the perfectibility of the human mind, seemed to indicate the necessity of a more extended sphere of action. Discontent and uneasiness were everywhere apparent. The faculties of man had begun to dwindle for want of scope, and the happiness of society required new and more copious contributions.

I reasoned with myself as follows: A bountiful Providence provides food for the appetite which it creates; therefore, the desire of mankind for a greater world to bustle in, manifested by their dissatisfaction with the one which they possess, is sufficient evidence that the means of gratification are provided. And who can doubt but that this is the time to find the means of satisfying so general a desire?

A great obstacle presented itself at the outset. The aid of steam in the navigation of my ship, was necessary to render my enterprise safe and expeditious against the adverse circumstances which I was sure to meet. But steam vessels were adapted only to smooth water. Every attempt to employ them upon the ocean had been unsuccessful. I foresaw that I must have a vessel capable of encountering severe gales in a dense atmosphere, of being rapidly impelled against strong currents, both of wind and water, and of surmounting, without harm, the impetuous tides, and resisting the violent winds to be expected in the polar seas. Moreover, she must be of such strength as to sustain the shock of floating ice, or of taking the ground; and of such capacity as to contain fuel and provisions for at least fifty men for three years, with apartments from which the external air could be excluded, and which might be artificially warmed during the rigours of a polar winter.

But he whose soul is fired with the true spirit of discovery, is not to be dismayed. I saw the end, and instantly began to use the means of attaining it. I caused a steam vessel of 400 tons to be constructed with double frames; the timbers being inclined from a perpendicular about 45 degrees; so that the outer set crossed the others at right angles. The timbers were let into each other to the depth of three inches and were secured by powerful bolts. This structure of massive grating was incalculably firmer than the frame of a ship could possibly be made upon the ordinary plan.

The bottom was covered with four-inch plank, over which, after they were fastened and caulked, a layer of three-inch plank was put on; and the whole was sheathed with copper of unusual thickness.

I remembered the misfortune of the discoverer Sinbad, whose ship, when he approached the magnetic mountain, fell to pieces, in consequence of the iron being all drawn out of it. To guard against a similar disaster, I fastened my vessel first with treenails, and then throughout with copper bolts firmly rivetted and clenched. To obviate the dangers of exposed and upright paddles, I built her with double topsides for a space of thirty feet. Within this space the inner frames sloped in from the bends, on an angle of 45 degrees, and were covered and finished, in all respects, like the sides of a common ship. The outer work was carried up in the usual manner, so that the aperture was not apparent to external observation. Through this outer side a longitudinal port was cut, 30 feet long and 3 feet wide, for the paddles to play through obliquely, like the fins of a seal. The nave of the wheel was two feet within the sill of the port, between the double walls, and supported by both of them. The blades of the paddles, made of the best ash timber, and firmly cooked and riveted together, were fitted into sockets in the nave; whence they could be easily unshipped for the purpose of closing the ports in bad weather, and rendering the vessel perfectly secure, with the paddles inboard. The shaft by which the power of steam was communicated to the paddles, passed through the inner side of the ship only, so that water could not be forced into the ship, even in the roughest weather, when the ports were closed. The inconvenience caused by the rolling of a vessel with upright wheels, was avoided by the obliquity of my paddles; the ship never rolling so much as to bring them to a perpendicular or dip the nave to which they were fastened. To avoid accidents from fire, I built beneath and on the sides of the furnace and boiler of the engine, two narrow cisterns, perfectly tight, and of incombustible materials. These were kept constantly filled with the wastewater of the engine, which was allowed to escape only by a spout at the top. No fire was permitted out of this enclosure. The economy of fuel, which was necessary from the length of the voyage, and from the emergencies which might happen, obliged me to adopt all the means of motion in my power. I therefore rigged my vessel as a ketch, with one large mast, and a long sliding topmast, which could be easily launched or sent up by the assistance of the engine; and a small mast abaft fitted to be struck at pleasure.

Having thus constructed a vessel which possessed the qualities most essential to my purpose, I finished the interior in such manner as I judged best calculated to render myself and people comfortable during the voyage. I took care to have one apartment large enough to contain all my crew. This was situated next to the furnace chamber, and had communication with it, by means of a tight covered passage. By a tube from the furnace, heated air could be conveyed to this apartment, and steam from the boiler by another tube, should the state of the air at any time require it.

Confident that, with this vessel, I could reach any place to which there was a passage by water, whether on the external or internal world, I named her the Explorer.

I furnished her with abundant stores for three years; among which were large supplies of dried and preserved vegetables and fruits, pickles, acids, and other antiscorbutic. The room not occupied by water and provisions, was filled with coal. Thinking I might meet with regions where none, but salt water could be had, from land, sea, or clouds, I took on board one of Youle's cambouses for converting salt into fresh water. Besides the best of cables, both of iron and hemp, and an extra supply of common and ice anchors, I failed not to provide one launch, as large as could be carried on deck, and four whale boats.

My next care was to select my officers and crew from among the most skillful, temperate, and orderly mariners I could find; whom I shipped for a sealing voyage in the South Seas, having a clause in the articles authorizing me to cruize and seek for seal wherever I might judge proper, for the term of three years. The crew consisted of 4 mates, 1 boatswain, 1 boatswain's mate, 3 engineers, 4 carpenters, 3 blacksmiths, 2 coopers, and 32 seamen: in all, 50 men, besides myself. In addition to a portable forge, and materials for repairing any damage which might happen to the engine, I took, on the suggestion of the chief blacksmith, duplicates of such parts of the engine as were most liable to fail. Of nautical instruments, chronometers, and books treating upon matters in any way connected with my object, I provided liberally. Least of all, did I omit Symmes's Memoirs, and printed Lectures. Finally, having completed my arrangements, and settled all my affairs, I took leave of my wife and children, whom, as I had no particular friends, I left to the humanity and kindness of the world, and set sail on the 1st day of August 1817.

CHAPTER II.

The Author arrives at the Falkland Islands—Describes West Point Island, and States harbor—Visits the city of the Gentoo Penguins on the Grand Jason—Gives some account of the polity and habits of those civilized amphibia—Sails for South Georgia.

I soon had cause to congratulate myself on my ingenuity. My fin paddles worked to admiration. When the wind failed, I could, by setting the engine in motion, propel my vessel at the rate of 12 knots per hour; and with a favorable wind, and under a press both of canvass and steam, found it easy to drive her at the rate of 16 knots.

With such advantages, there was no necessity of going the roundabout passage to gain the trade wind. I therefore stood straight for Cape St. Roque. Whether I did or did not see a flying fish, catch a dolphin, or observe a black whirling cloud called a waterspout, is of very little importance to the world. On the sixteenth day after leaving port, we saw the land of Cape St. Roque, in South America, and on the twenty-fourth, anchored in the harbor of Rio de Janeiro, having experienced the usual changes of wind and weather, and discovered that air and water are much the same elements, and are governed by much the same laws, at sea as on shore.

I entered this harbor under sail, with the paddle ports closed, that no suspicion might be excited; my object in calling at this place being only to provide myself with livestock and fruits. I took on board. two fine horses, four mules, two cows, with calves, a parcel of pigs, sheep, and goats, with a quantity of fruit and vegetables; and, on the 26th of August, sailed again.

On the 4th of September, we entered the harbor of West Point, Falkland Islands. Here 1 had determined to pass a month for the benefit of my health, which a short passage by water had not completely restored, from the debility occasioned by the vexations and anxieties of business in those retrograde times, and the pernicious habits of living, common among civilized men, upon food rendered palatable by a skillful admixture of poisons. These Islands being incontestably in the healthiest region of the globe, I believed that, by a short stay amongst them, I should regain the firm health so necessary to a man who undertakes great things; and at the same time, by employing my people in sealing, teach them how to manage the boats, to land through a surf, and to execute all the difficult and dangerous operations, incident to the occupation of explorers of unknown shores. At the same time, I should be pursuing the ostensible object of my voyage; a matter very necessary to be kept in view, for my people were engaged on shares of what should be obtained by their industry.

The first day was devoted partly to preparations fin. a sealing excursion to the Jason Islands, and partly to recreation. West Point Island abounds with hogs and goats, the hunting of which is both pleasant for exercise, and profitable by supplying excellent food. Here are no tangled forests to embarrass the sportsman, nor bushes or briars to annoy his clothes or his flesh. Neither are their mats, moschetos, sand-flies, snakes, scorpions, or other reptiles, to render every step dangerous or painful. Near the shore, which is fringed with granite rocks, a border of tussoc extends around the island, like a belt, of from one-eighth to a quarter of a mile in width. The tussoc flag grows from the top of a bog formed apparently by the roots of the plants which had flourished and decayed on the spot for many successive years. The bogs are, usually, three to five feet in height, and one to three feet in diameter. The substance of them resembles cork, though it is less compact. They stand irregularly one to two feet asunder, so as to afford convenient room for a passage between them, in every direction, over a foundation of much the same substance as the bogs themselves, which is usually quite dry. The dry white tussoc grass of preceding years hangs round the top of the bog like a broad frill; while the fresh green growth, which waves over the top like a tuft of lofty feathers, gives the whole, when viewed from a distance, the aspect of an extensive field of indian corn. The root of the fresh tussoc is pleasant to the palate, being much like the meat of a chestnut, and it affords an abundance of excellent feed to the hogs, that enjoy an Elysium here. Within this border of tussoc, and from it to the steep ascent of the mountains, a region of grass intervenes, which has the appearance of a rich upland meadow. It grows about knee high and extends as far as the rise of the land is moderate. Beyond it, short mountain grass and a few heath plants are found contending with fragments of granite, and with the polar blasts to which the lofty summit of the mountain is exposed. It was delightful, after a confinement on shipboard, to ramble over this sequestered and pleasant scene; to chase the wild hogs from their tussoc covert to the rising grounds, where they were sure victims of the spear or the bullet; and to invade from above the retreats of the gigantic albatross, in the cliffs of perpendicular rocks, a thousand feet above the sea.

On the second day we landed a sealing party of thirty men, under the command of Mr. Boneto, chief mate, on the Jason Islands, which are similar in their formation to those I have described. Intending to join this party myself with the launch and being apprehensive that if I left the Explorer in West Point harbor, with but a few men on board, some Patriot pirate might look into that much frequented place and tempted by the value of my vessel and her defenseless situation, deem it patriotic. to take her away to aid the cause of liberty and leave me to explore my way home in my boats, I proceeded with her to the deep and intricate inlet of the sea, called States harbor. This spacious, convenient, and secure harbor, second to none on the face of the globe, is one of the indications that Providence formed this group of Islands for the abode of an enlightened and maritime people. Front a spacious and deep bay, in which the whole navy of Britain might moor in safety, a cove jets into the land on the left; and on one side of the cove there is an opening through the land like a dock-gate, with perpendicular sides of solid rock, against which a ship of the line might lie with safety, as against a pier. Passing through this opening. a harbor is found, extending at right angles with the passage nearly two miles in length, and about one-eighth of a mile in width. At one end of this interior basin, a large stream of fresh water empties into it; at the mouth of which fine fish in great quantities are easily taken in the spring, and on its banks, as also on those of numerous smaller streams, celery of an excellent kind grows spontaneously. The shores of this basin rise with a very gentle ascent. They are not exposed to the winds of the open ocean and are not much encumbered with tussoc. There is no high land near thousands of acres, well-watered and covered with grass fit for hay, exhibit the prospect of a succession of well-cultivated meadows. There are plenty of hogs on the island which forms this harbour. Geese, as good as our wild geese, are very abundant. We caught them with ease, and in great plenty.

What a delightful situation these islands offer, for a virtuous, enlightened, and industrious community! Nearly four hundred islands, one of which is some hundreds of miles in extent, situated in the most temperate climate of the globe, where the air is always salubrious, heat never oppressive, cold never severe, the ground never frozen, and the heaviest snow of no more than two or three days duration on the ground; with a soil capable of affording, by cultivation, all the useful products of the temperate zone; a location convenient for the prosecution of the whale, seal, and other fisheries; with innumerable harbors for the accommodation of shipping; with everything calculated to make them the most desirable residence for man, these islands remain uninhabited, and lonely deserts.

The fine health which those who stop here for a season invariably enjoy, the appetite they acquire, the activity with which they exert themselves, these are the evidence of an invigorating and salubrious climate. Here are no debilitating heats to enervate, nor frosts to benumb the faculties, and make it half the business of life to keep the body comfortable. A people born and educated in such a country might be expected to partake its characteristics; to have minds solid and profound, like the granite frame of their mountains, and clear as the ocean which surrounds them; vigorous, yet temperate like the climate; sufficient in all things, and without extremes.

Having seen my vessel safely moored, I left her in charge of Mr. Albicore, the second mate, with strict orders not to permit either fire or candle to be used on board in my absence. I caused a cook house to be erected on shore, and left five men with Mr. Albicore, to fill up the water-casks, catch and cure fish, make the necessary repairs to the rigging, and put the vessel in perfect order against my return. With the remainder of the officers and men, in the launch and one whale boat, I made a harbor at West Point Island early the first day, and at the close of the second joined the party under Mr. Boneto, on the Grand Jason. I found that Boneto had made good use of his time, having cleared this island and all the neighboring keys and shores to which he could prudently go with open whale boats, of the few seal which could be found. There was but here and there a seal to be seen, excepting on some points of land, which on account of the surf were nearly inaccessible. The frequent visits of sealers from the United States had either destroyed or frightened most of them away. This gave me no uneasiness, for I expected it when I planned my voyage. I concurred in the opinion published by Capt. Symmes, that seals, whales, and mackerel, come from the internal world through the openings at the poles; and was aware of the fact, that the nearer we approach those openings, the more abundant do we find seals and whaled I felt perfectly satisfied that I had only to find an opening in the "icy hoop," through which I could dash with my vessel, to discover a region where seals could be taken as fast as they could be stripped and cured. I therefore employed myself chiefly in procuring comforts for my people, and in studying the habits and propensities of those amphibious animals which might be supposed to have communication with the internal world, whither I was ambitious to find my way.

A colony of Gentoo Penguins, on the borders of the south-east cove of Grand Jason, first attracted my attention. Their city stands on a beautiful level spot, a short distance from the water. Every pair of Penguins has a separate establishment built of earth, stones, and sticks, of about two feet elevation, and eighteen inches diameter; on the top of which is their nest. There are some thousands of these stands arranged in regular order, with an open square in the centre, regular streets between the ranges of nests, and a broad avenue leading from the square towards the places of landing and diving. This avenue, a short distance from the settlement, divides into two broad paths; one leading to the diving place, which is a perpendicular rock in deep water, and the other to the landing place, which is a sloping rock of easy access from the sea.

It being the egg season, which soon passes away, I determined to make it hold out, if practicable, until the lime of our departure, that we might have a stock of fresh eggs to take with us. Remembering that our barn-yard fowls continue to deposit eggs as long as but one is daily left in the nest, I adopted that plan with the Penguins, and stationed Jack Whiffle, boatswains-mate, with three assistants, to remove the eggs daily, and stack them; keeping an account of the several stacks, that we might take our supply from those lest gathered. This was no trifling job. The nests were so numerous that it was a hard morning's work for four men to visit them all and take an egg from each in defiance of the lawful proprietor, who always defended his property to the best of his ability, and never forsook the stand, through fear, but maintained possession until pushed off. The plan answered my expectations: The Penguins continued to supply eggs in place of those that were removed, until our departure, when we took with us barrels of them packed in salt.

These Gentoo Penguins are amphibious birds, nearly two feet high when standing erect. Their bodies are somewhat larger than those of geese, and well-proportioned throughout; their necks being just long enough to look well. In place of wings, they have fins for swimming, and their feet are equally well adapted to the land and water. Their covering is very short feathers, closely and firmly set in a thick skin. Their backs, fins, feet, and legs, are black; the rest of their bodies pure white; they walk bolt upright, with a firm step like a grenadier, and have the appearance, when formed in squadrons, of soldiers, in a uniform of black coats, white underdress, and black gaiters. From the attentive observations of Jack Whiffle, I obtained the following particulars of their habits and polity:

At the time of full sea, one half of the Penguins assemble in the centre-square, where they parade in regular order. They then march off, two abreast, and in close order, preceded by a leader, to the diving-place.

They dive into the sea in succession, as they arrive, and swim off to feed on kelp, rockweed, small fish, and other marine productions. During their absence, the other half remain stationary upon their stands, keeping watch; occasional short visits by some few of them to their nearest neighbors, being the only deviation from strict duty in this particular, that is allowed. If anyone strays far from his station, or shows a disposition to go out to feed, he is pecked and driven back by the others. At the turn of the tide, those that are out collect about the landing-place; some sporting in the water, leaping and diving with great dexterity; others lounging upon the shore, apparently admiring themselves and each other, like our fashionable belles and dandies in Broadway. When the leader lands, they form in regular order, march to the square, in the same manner as they left it, divide into squadrons, and file off to their respective stations to relieve guard. As soon as those returned from feeding mount the stands, the others leap off and repair to the square. When collected, they form, and march off to the diving place in the manner before described, to take their tide of feeding and recreation. Thus, they occupy the day; each having the benefit of a full tide, and each doing his share of domestic duties.

At night all are assembled in the city, and each stand is crowned with two of these exemplary birds.

The contemplation of these orderly, discreet, and beautiful amphibia, afforded me much pleasure, and gave rise to many delightful anticipations. It appeared certain to me that they, in common with seals, whales, and mackerel, were visitors from the internal world through the southern opening, which they were admirably formed to pass and repass; for they moved with great facility, in the water, and could exist under it as well as fish. On land they walked with as much ease as men, and in the same erect posture. It occurred to me that a world, in which the brute creation was so neatly formed, so polished in their manners, so social in their habits, and so quiet and well behaved, must, if men existed in it, be the abode of a race perfect in their kind. I had no apprehension of the air being unhealthy in the internal world, as suggested by Capt. Symmes, because the climate in which these visitors are found in the greatest numbers is the healthiest of the external world, which indicates that they are accustomed to good air, or they would not affect this salubrious region.

Again, I had observed all these amphibia to be of a remarkably gentle and harmless disposition. The sea-lion, sea-elephant, and common seal, together with the king-penguin, the Gentoo, macaroni, and jackass-penguin, all of different habits, yet obviously of the same origin, accommodated themselves on the same island, fed in the same sea, and on the same food; without interfering with, and without ever being observed to offer violence to each other; from which I inferred that the inhabitants of the internal world, influenced by the same causes, must be of a remarkably pacific, and gentle disposition.

October had arrived, and I grew impatient of further delay. The sun was already pouring its rays of light and heat a constant stream upon the south pole. The season for active research in that region was come and would soon he past. I directed Mr. Boneto to collect the skins which had been taken, at a convenient place on Grand Jason, and returned to the Explorer. I found everything at States harbor as it should be. Mr. Albicore was an excellent officer. He took care to understand my orders, and to obey them implicitly. The launch was immediately hoisted in, and at dawn of day the following morning we sailed from that port, took in Boneto's party, with near two thousand seal skins, and bore up for South Georgia.

CHAPTER III.

The author passes South Georgia, and proceeds in search of Sandwich land—States to his officers and men his reasons for believing in the existence of great bodies of land within the Antarctic circle, and for the opinion that the polar region is subject to great heat in summer. —Crew mutiny at the instigation of Mr. Slim, third mate—Happy discovery of a southern continent, which, at the unanimous and earnest solicitation of his officers and men, lie names Seaborn's land.

On the 10th of October we approached the principal harbor of South Georgia, which I had no intention to enter unless there was an appearance of an unusual abundance of seal on the coast. When near the harbor we discovered two ships lying there with their topmasts struck. This was evidence that there was no chance for us in that quarter. I now told my officers and people that I thought it useless to contend with those already in possession of the island for the few seal it could afford and thought it most advisable to proceed in search of Sandwich land; where, no doubt, we could speedily obtain a full cargo of skins, if we could find it, of which I expressed great confidence. I strengthened their hopes by assuring them that there was no doubt in my mind of the existence of extensive bodies of land within the antarctic circle, which quarter had scarcely been looked into by Christian navigators, and that my opinion was founded upon the fact that Cook, and other navigators, had seen large bodies of ice in latitude 70° to 71° south. This fact, I said, indicated the existence of land, because ice could not form in a deep salt sea uninterrupted by land, and agitated by the violent winds and currents of the polar region. I urged that we had but to persevere in our research in high southern latitudes, to make sure of finding laud, which would yield us ample fortunes, for all southern islands, when first discovered, were found to abound in seal. Mr.

Slim, the third mate, expressed some apprehension, that great danger might be encountered in high southern latitudes; that if we found land, the ice might close upon us and prevent our return to our country, as it once served a colony in Greenland. I was not much pleased with this. I have no patience with an officer who suggests doubts and difficulties when I have a grand project in view. I marked him, but at the same time pretended to listen to his observations, as objections of great weight, and then proceeded to remove them from the minds of the officers and people, by advancing the following reasons for my belief that the supposition of extreme cold at the pole was altogether gratuitous.

1st. We know that the rays of the sun., uninfluenced by the atmosphere, would rest upon the pole for six successive months.

2d. That a dense medium refracts or bends the rays of the sun.

3d. That the amount of that refraction depends upon the extent of the dense medium through which it has to pass.

4th. That at the pole, the rays of the sun coming to it in a very oblique direction, must necessarily pass through our atmosphere a greater distance than on any other part of this globe, and consequently must there be refracted in a greater degree than elsewhere. Hence I inferred, that in consequence of this refraction, and of its increase in proportion to the obliquity of the direction of the rays, the sun when in the plane of the equator, must appear to an observer at the poles to be some degrees above the horizon, and that the sun must recede to the north of the equator at least five or six degrees of declination, before it would become invisible at the south pole: therefore, as it takes fifteen days to increase the sun's declination five degrees, it must be visible at the poles one month longer, on account of the refraction, than it would be without it. This conclusion is corroborated by the testimony of Barentz, a Dutchman, who wintered in Nova Zembla. He found the sun to rise, in latitude 76°, fifteen days sooner than was expected by astronomical calculations. This will give the polar region seven months constant sunshine; think of that, my shipmates, said I, seven months constant day, with a continual stream of light and heat pouring upon the same spot, without any interval of night to cool the earth and air. I think if we can but find our way to the polar region, we shall be in much more danger of being roasted alive, than of being frozen to death. But, my lads, what Yankee sailor would hesitate to expose himself to be roasted or frozen alive to accomplish that which the British tars have endeavoured in vain to do? Three hearty cheers put an end to the debate. We bore up for *Sandwich* land, not that I had any belief in the existence of any *such* land, for I had always been of opinion, that the English placed this supposed land on their charts as an English discovery, stretching it along from the polar seas to latitude 57° south, that they might, whenever any land should be discovered in that unexplored quarter, have a pretense for laying claim to it as a British discovery.

We had a fine gale from the S. W. and made rapid progress to the S. E. under canvass. Although the most perfect satisfaction with the course I had determined on appeared to prevail throughout the ship's company, Mr. Slim came to me in my cabin, when relieved from his watch on deck, and told me, that, however satisfactory my account of the matter might have been to the other officers and the crew, it was not satisfactory to his mind; and he should be glad to be informed how I accounted for the vast bodies of ice which had invariably stopped the progress of navigators in high latitudes, if my notions of great heat at the poles were correct? "Take a chair, Mr. Slim, and we will talk about it. In the first place, we have no account of any navigator having sailed to a higher southern latitude than 71°, and 82° appears, from the most authentic accounts, to be the highest northern latitude that has been visited. Navigators to these high latitudes have always found ice between the parallels of 70° and 80°, which space that profound philosopher, John Cleve Symmes, denominates the 'icy hoop.' It is true he has not taken the trouble to explain to the world, in a satisfactory manner, why and wherefore this narrow strip of ice should exist in that region, which omission, I judge, must have arisen from the circumstance of its being obvious to his capacious mind, that such a 'hoop' must necessarily exist, 'according to the laws of matter and motion.' The causes of it appearing to him perfectly simple, he could not suppose it necessary to state them to 'the most enlightened people on the face of the globe.' Now, sir, I will explain the matter to you. At the pole, that is, ninety degrees from the equator, there is seven months summer, without any interval of night, as I stated on deck; and when the sun has twenty-three and a half degrees of south declination, its rays must strike the pole, allowing but three degrees for the effect of refraction, on an angle of 26½° with the plane of the horizon, and must appear nearly as high as in Scotland in the months of March and September. It is true it does not continue at this extreme declination for any great length of time. On the other hand, it does not recede so far as to withdraw its rays from the pole for a single hour during seven months of the year. This we know; and you can imagine, from the effect of a March sun, which in your country, Mr. Slim, loosens the icy fetters of winter, although withdrawn one half of the time, what must be its effect when exerting its influence for months without any interruption? Now in latitude 70°, with the exception of a few days, there is an interval of night the year round. In the winter months the climate cannot differ much from that of the pole. The cold is then no doubt severe, and forms ice in both those positions. In the early part of summer, that is, September, October, and November, there is at the pole a steady blaze of heat and light, which must melt the ice accumulated in winter, by causing a constant thaw. This sunshine continues at the pole till the 1st of April and prevents the forming of ice until that time. But at 70°, there is, through most of these months, a short period of night, sufficient for the atmosphere to cool. This will be more obvious, if we consider the powerful influence of the ice, during this absence of the sun's rays, and remember the great change of temperature which occurs in our climate immediately after sunset at the close of a sunny day in February or March. This interval of night in latitude 70°, counteracts most of the effects of the sun's heat in the daytime. Nearly as much ice forms in the night as is thawed during the day. This accounts for the 'icy hoop.' There is not summer enough to dissipate the ice of winter; while at the pole there is summer enough to dissolve a globe of ice."

"But, sir," rejoined Mr. Slim, "if this 'icy hoop' exists, how do you expect to pass it? or, if it is impassable, what use is there in encountering the risk of navigating unknown and dangerous seas, in a high and boisterous latitude?"

"I mean, sir, to ascertain whether it be passable or not. I think it probable that the influence of the summer heat may so far weaken it as to admit of broad openings being formed by the pressure of wind or currents, and if I can find an opening of but a mile wide, I shall dash through it, at all hazards."

"And a pretty condition we shall be in, Capt. Seaborn, if the ice closes the passage after we have dashed through it!" replied Mr. Slim. "We shipped with you, sir, for a sealing voyage; not for a voyage of discovery."

"You will please remember, Mr. Slim, that I am expressly authorized by the articles, to cruize and seek for seals wheresoever I may judge expedient and proper, and that any opposition to my authority will involve the forfeiture of your share—recollect that, Mr. Slim."

"I do recollect that, sir; but at the same time, I know, Capt. Seaborn, that you have no right to hazard all our lives, by running into dangers, greater than were ever encountered by human beings, to gratify your mad passion for discovery, instead of pursuing the interest of all concerned, by endeavoring to find seals in the usual manner. How will you justify yourself to the world, to our families, or to your own conscience, if we should, after effecting a passage through this 'icy hoop' you speak of; find it closed against our return, and be thus for ever lost to our wives, our children, and society? We must in such a case all perish, and our blood would be upon your head." A plague upon your lean carcass, thought I, how am I to answer so many impertinent questions. I could not tell him of my belief of open poles, affording a practicable passage to the internal world, and of my confident expectations of finding comfortable winter quarters inside; for he would take that as evidence of my being insane, and by means of it persuade the crew to dispossess me of my command, and confine me to my cabin for the remainder of the voyage. After knitting my brows a short time, I replied, "Mr. Slim, you are a sufficiently capable officer, and can get through with your duty well enough when you choose to do it, but you don't know everything; your mind is too dense to admit the rays of intelligence. I would have you to know, Sir, that I command this ship, and am not to be thwarted or dictated to by any man. I have noticed your rebellious spirit; now mark me, Sir, so sure as I have any more of your opposition to my will, or hear any more of your murmuring; the moment I detect you in uttering one discouraging word in the hearing of any of my officers or men,— I will confine you, and carry you home in irons, to take your trial for conspiring to make a revolt in the ship, which is death by the law; remember that, and go to your duty, Sir."

Slim had some prudence and was a great lover of pelf; he did not relish the idea of forfeiting his share; he kept his tongue between his teeth; but his lank, expressive features spoke horrible things. This comes of taking more officers than there is duty for, thought I, as he left the cabin; that fellow will give me trouble enough before I get rid of him; there is nothing like constant hard work to keep men out of mischief. But I had not much time for reflection, for Will Mackerel, my fourth mate, whose birth adjoined my cabin, had overheard all that passed in my interview with Slim, and as soon as he was out of the way bolted into my cabin, without much ceremony. Will was a hearty, frank, thorough bred sailor; doffing his hat to his commander was the only point of etiquette he was acquainted with, and he thought it degrading to perform that ceremony to any other person. Will reverenced his commander when he found him to be a good sailor, a skillful navigator, and a kindhearted man. He commenced with, "Captain Seaborn, that fellow's insolence is insufferable; he has spoken more mutiny to your face, in your own cabin, in ten minutes. than all the rest of the ship's company would dare to think of, in the forecastle, the whole voyage. I would not give a rope yarn for a sailor who would not go wherever the captain had courage to lead the way. I would not put up with it; there is but one Slim in the ship, and we'll heave him overboard, if you say the word; at least, I'd clap the ruffles * on him, and keep him out of harm's way the rest of the voyage."

Whether it was honest indignation that prompted Will's advice, or whether some little desire for Slim's birth, to which he would be promoted of course if Slim were cashiered, had its influence, I did not stop to ascertain. I told Will, to be quiet, to say nothing of what had passed between me and Mr. Slim, but to observe him closely, and let me know if he detected him in endeavours to corrupt the crew.

We made rapid progress, and were soon in the latitude of Sandwich land, as laid down in the charts, where we met with nothing but clear blue ocean. I hauled up S. S. E., true course, and stood on as far as 68° South, making the best use of my time by daylight, and drifting back upon my track during the short interval of night. On the 2d Nov. in lat. 68½, we met with ice in detached fields; and had strong gales from S. W. with raw, drizzly weather. I edged away to the eastward, intending to keep near the ice, and hauled to the southward, when a clear sea would permit. The first day, we kept the 'blink of the ice' * in sight and found it to trend nearly East and West. Made no southing this day. The second, we were enabled to haul up S. E. and by E. and continued this course without nearing the ice. The following day, hauled up S. E., set the engine in motion, and made rapid way; we observed this day at noon, in 75° 22′ S. I was elated with the prospect of reaching a much higher southern latitude than any former navigator had been able to gain, and pushed on as fast as canvas and steam could drive my vessel.

We had no interval of night in this high latitude, the sun's declination being 15° S. After running on this course 24 hours, we lost sight of the ice entirely, and thinking it most prudent to keep close under the lee of the ice to windward, that in case of a hard gale we might have smooth water, I steered due South. We observed this day, 5th Nov. in 78° 10′, with cold, raw, disagreeable weather.

I had observed Slim moving about the ship, like an uneasy spirit compelled to revisit this troubled world, often whispering to the men, and frequently visiting the forecastle. When I came on deck after dinner, the whole ship's company came aft, with Slim at their head, who in their behalf told me, that the crew had determined to go no further with me into this region of ice. Will Mackerel, who was on the quarter deck, spoke up with great passion, and asked Slim if he meant to head a mutiny? adding, that if such was the case, he would let him see that he was a man to stand by his commander. He then called upon those who were of his mettle, to come over to the starboard side; which some few did, while some took their stand amidships, that they might go either way, as circumstances should dictate. The greater number, however, remained with Slim. There was a sad uproar for a short time, everyone having something to say, and to enforce with an oath. Even the man on the lookout at the mast head came down from his station to take a part in the affair.

While this war of words was going on, Mr. Boneto, who was below, hearing high words on deck, came up with his hanger and pistols; and the steward brought me mine, but I ordered him to put them up again, saying, if the men will not listen to reason, we will give up the voyage. The truth was, I felt sensible that had I been possessed of my pistols at the outset, I should certainly have shot Mr. Slim; but at this time the irritation of the first impulse had subsided a little. I had had time to cool. Mr. Albicore was standing by my side, as mute as a fish, waiting for orders. The boatswain Jack Whiffle his mate, and a number of the best men, had joined Will Mackerel's party; while those who adhered to Slim were the poorest seamen, and most timid men in the ship, though at the same time the most noisy.

How the matter would have terminated but for a lucky occurrence is doubtful. The vessel was running on her course during this contest, with no one on the lookout: a splash in the water, close aboard to windward, drew my attention that way; it was a seal. At the same moment I observed the water to be discolored, and instantly ordered the engine to be stopped, and a cast of the lead to be made. Some of the faithful hastened to execute this order under the direction of Albicore and Will Mackerel: but Slim and his malcontents kept up their vociferation, Slim telling them that it was only a maneuver of mine to divert them from their purpose.

While this was going on, I swept the horizon with my spyglass, and soon discovered in the S. W. directly to windward, a low range of broken land. The moment I fixed my glass upon it, every eye was turned in that direction: some sprang into the rigging, some ran to the mast head, and the joyful cry of land ho! land! dispelled the mutinous disposition of the crew.

Sixty-five fathoms, soft ooze, was the report of soundings: a delightful indication of an extensive body of land, with large rivers depositing their sediment on the bottom of the deep. We soon approached and observed the coast to range about S. E. and N. W. as far as the eye could reach from the mast head. I called the attention of my officers to this circumstance and observed to them that the broad opening which we had found in the 'icy hoop,' could now be easily accounted for. We had noticed that the prevailing wind was from the S. W. with strong gales, the influence of which was continually forcing the ice to the eastward; but this body of land, ranging from the S. E. to the N. W. stopped the ice to the westward of it, while that to the eastward was driven away, leaving a clear passage to leeward of the land. From the westerly winds prevailing all the year round, this must always be the case, unless the 'immutable laws of matter and Motion,' and the relation between cause and effect should be changed.

Mr. Slim, who had been leaning over the rail with his back towards me during my discourse, now turned upon me, with "well said, captain, that is the best reasoning I have heard from you yet, —I understand that." The truth was, we were now well in with the land, and the appearance of vast numbers of seal in the water and upon the shore; gave a prospect of a splendid voyage, and excited Slim's cupidity, and his apprehension, for the safety of his share, which he was aware he had jeopardized by his conduct.

I was in excellent good humor and told Slim I would overlook what had passed; I could do no less, at a moment when a kind providence was favoring my enterprise beyond my hopes, notwithstanding, my numerous transgressions, without evincing an ungrateful and malicious spirit. The utmost joy prevailed throughout the ship's company; even Slim's livid countenance was distorted with an unusual grin. Slim, was not without shrewdness, and occasionally he pretended to be very religious; but he had a double allowance of native selfishness and worshipped with heartfelt devotion no other god but gold. With his misconduct forgiven, and a prospect of gain which surpassed his most sanguine expectations, he felt emotions as much like those of happiness, as such a compound of evil passions could be supposed to feel.

When near the land, I observed it to be in general very low; there was scarcely any appearance of elevated spots, and no high hills or mountains could be seen. From the rugged appearance of the coast, I judged that there were deep indentations, affording numerous and convenient harbors, but in this I was mistaken. What we had taken for the coast, proved to be a succession of islands, with a broad sound between them and the mainland, which later had a straight, unbroken shore. Deep water, and a very rapid current or tide, rendered it unsafe to anchor amongst the islands; we therefore continued to coast along the main shore in search of a harbor for several hours. The shore in this place was not elevated more than 30 to 40 feet above the level of the sea. It was skirted with tussoc, which, from the very gradual rise of the land, hid all the interior from our view, except a few moderate elevations far distant.

At 6 P. M. the appearance of a wide bay induced me to send off the boat to examine for anchorage. At 10 they returned, with the information that the bay afforded good shelter with soft ground, but was rather objectionable as a harbour, in so high a latitude, on account of its being full four miles wide, and very deep. I determined to run in and anchor, until a more secure port could be found; and having despatched two boats ahead to report the soundings by signal, stood into the bay, and at 12 o'clock P. M. anchored in 10 fathoms, soft mud, the two capes of the bay in one S. S. E. about one league, the western shore one mile distant. Although it was midnight according to our reckoning, we had a bright sunshine, the sun appearing ten degrees above the horizon.

This land having been first seen by myself, my officers and men united in calling it after my name and expressed their wish that I would permit it to be so denominated; it was accordingly recorded in the ship's logbook by the name of Seaborn's Land.

I had much need of rest, having been almost constantly on deck for five days; and after ordering the deck watch to get the boats out, and prepare everything for an excursion, I retired to my cabin, and was soon fast asleep.

CHAPTER IV.

The Author in great peril, from the vast rise and fall of the tide in the polar sea—Brief account of his observations at Seaborn's Land—He takes formal possession of the country, in the manner usual in such cases, in the name and on behalf of the United States—Leaves a sealing party on one of the islands near the coast, and proceeds to the south, to extend his discoveries.

I had slept some hours, when I was awakened by Mr. Boneto's order, and informed that the land appeared to rise very much. I went immediately on deck and was astonished to see the land appear more than three times as high as when we came to anchor. I at first attempted to account for it by supposing some change in the atmosphere which caused the land to loom; but was soon undeceived. One of the seamen called out that there was a shoal even with the water close by. The lead was immediately cast to see if the ship was driving, and but two fathoms water were found alongside. In half an hour more we were high and dry. Such was the astonishing rise and fall of the tide in this high latitude! The bay, which had twenty fathoms water in the centre at full sea, and ten fathoms a mile from the shore, was almost entirely emptied; a small channel in the middle, not more than half a mile wide, being all that was not left quite hare. There was no immediate inconvenience to be apprehended from this circumstance; but I was aware, that a tide that fell 70 or 80 feet perpendicular, must return in a bore with prodigious violence, and was under more apprehension of the consequences, than at any other period of my voyage. I however concealed my fears from my officers and people, who were much amused with the circumstance, and my apparent vexation at finding my vessel high and dry on a mud bank, near the south pole. My greatest fear was, that the tide might come in in a bore thirty or forty feet high, and, striking the vessel as she lay aground, tumble her over and dash her to pieces, no frame of timber being sufficient to withstand such a shock. Happily, the stream of the ebb tide had left us exactly stern to the flood. I ordered the boats to be hoisted in and secured, and the anchors to be taken up, fastened in the dead lights, put everything below that was moveable, directed the men to provide themselves with strong lashings, and ordered the engineer to raise a head of steam, and have the engine in readiness for instant motion. Thus prepared, I awaited the return of the tide. It came in dun time; and now my officers and men, who had been so merry at my expense, evinced great consternation. The muscles of Slim's face were actually convulsed with terror at the sight of a wall of water, stretching quite across the bay, apparently 30 or 40 feet high, rolling towards us like a tremendous breaker, and with a roaring noise like thunder. To all appearance, it would break over our mast head, and consign us to one common grave. In mercy to the trembling Slim, I desired him to step below and bring me my pea jacket, well knowing he would not come up again until the danger was over. I then ordered the companion way and the hatches to be secured, directed my people to lash themselves fast, and quietly wait the result. Here, I must confess, I put up a silent prayer to Heaven, after a sailor's fashion, for preservation from the impending danger.

I have always found the fears and anticipation of danger to exceed the reality. When the bore approached us, the bottom came rather faster than the top, and its face was not quite perpendicular. The vessel was fairly afloat on the foot of the wave, before the main body of it struck her; and taking it square astern, she split and rose over it in the most beautiful manner, without sustaining the slightest injury. By backing with the paddles, we kept clear of the shore, on which the impetus of the wave would have driven us, and soon after anchored again in the middle of the bay in twenty fathoms water.

And here I would recommend to all navigators of the polar seas, to avoid anchoring in less than twenty fathoms, until they have accurately ascertained the rise and fall of the tides, at the full and change of the moon.

When the companion way was unbarred, Slim came up with my pea jacket, and coolly observed, he was glad there was no damage done, adding, "I was really afraid it might break our paddles." In consequence of this occurrence, I named this bay Take-in harbor.

We were occupied until noon, in returning things to their places, getting the boats out, and preparing for an excursion on chore. At noon I observed the altitude of the sun, and, after making accurate allowance for the refraction, found Take-in harbor to be in latitude 83° 3′ south. This was much further south than the distance run by log would make us, which I attributed to a strong current setting us rapidly in that direction; but this I soon found to be an error, and that the difference between the latitude by observation and dead reckoning, arose from the oblate form of the globe at the poles, lessening the degrees of latitude.

After dinner, I landed with a strong party, leaving the vessel in charge of Mr. Boneto. I took the horses and mules on shore, with provisions for a week, intending to march to the highest land we could find, to gain at once an extensive view of the coast and country. We landed on the south side of the bay and shaped our course. for a moderately elevated spot, which appeared to be the highest land, due south about ten miles distant. We found the shore much like that of the Falkland Islands, the only difference being that this was much more level and had greater extent of tussoc. After passing through a border of tussoc about three miles wide, we reached an open prairie country, with grass about four inches high. We were three hours in gaining the elevated spot, from which we were enabled to see the coast for a great distance on our left, and the sea along its border, studded with islands. On the right, we could see nothing but boundless prairie, with here and there a ridge like the one we were upon. To the south, in the horizon, appeared something like a hill, and to that I determined to go. Having taken some refreshments, we took up the line of march. Slim, who was with me, as I did not think it prudent to leave him on board, had been very docile until now: on finding me determined to push into the interior so great a distance, he became evidently uneasy. He dared not express his fears to me but took care that I should overhear him say to one of the men, "I hope the captain won't waste so much time in exploring this desert, that we shall be obliged to go away without a full cargo of skins, or run the risk of being obliged to winter here, so near the pole, where we should certainly all freeze to death, in spite of everything we could do." As this was a reasonable apprehension in the mind of an ignorant man, I endeavored to remove his fears by calling his attention to the tussoc grass and other plants, and asked him how they survived the winter, if the cold was so intense as he supposed? and advanced the opinion that wherever plants can sustain the cold of winter, and retain their vitality, man can exist, with the aid of good clothing and artificial heat.

A fatiguing march of 15 miles brought us to the hill, which we found to be the highest part of a ridge of moderate elevation running from the coast in a S. S. W. direction into the interior. We were amply compensated for our trouble in wading through the grass, as this eminence afforded an extensive view of the country in every direction. The S. E. side of this ridge broke off very abruptly, in some places perpendicularly and at its foot was a large and beautiful river, full a mile in width, flowing from the S. S. W. Beyond it was a prairie country, gently waving and rising into sloping hills in the distant horizon. Far up the river I could descry with my glass a few trees, towards which I felt a strong inclination to proceed; but being excessively fatigued, thought best to devote a few hours to refreshment. After a comfortable meal, and a sound nap of four hours, I descended the precipice to ascertain whether the river was an arm of the sea, or a freshwater stream. It proved to be pure potable water, and the existence of a continent near the south pole, was thus fully established.

I had not been long on the bank of this river, before I found cause to doubt the prudence of venturing thus far by land into an unknown country, in the appearance of fresh tracks of some huge land animal, which were larger than the bottom of a water bucket. Whether they were those of a white polar bear as big as an elephant, of a mammoth, or of some other enormous nondescript animal, I could not guess. I re-ascended the hill with all practicable expedition, collected my men, and hastened towards the ship as fast as possible.

We reached the ship after six hours constant marching, all completely tired out, our horses and mules being too feeble to travel, from long confinement on shipboard.

The discoveries I had already made were so far from satisfying my ambition, that my desire to push on and explore the internal world was more intense than ever. I was now convinced of the correctness of Capt. Symmes's theory, and of the practicability of sailing into the globe at the south pole, and of returning home by way of the north pole, if no land intervened to obstruct the passage. My first thought was to enter the river I had seen, and ascend to its source, which must necessarily be in the internal world; for if the poles were open, there was not room enough for a sufficient body of land to the south of 84 degrees, to maintain so mighty a river. But I abandoned this idea, on reflecting that by confining myself to this river, I should at best enter the internal world but a few hundred miles, while by entering on the open ocean, I should be able to visit every accessible part of it.

My first business was to make such arrangements as would satisfy my crew, and to ascertain the condition of the country in the immediate vicinity, therefore landed a sealing party of thirty men, under Mr. Boneto, assisted by Mr. Slim, on one of the islands, and proceeded with the Explorer to the mouth of the great river: We found the access to the river easy and safe; the chain of islands off the mouth of it broke the swell of the sea. Having ascertained its mouth to be in 83° 47′ south latitude, by observation, I proceeded up with two boats ahead, taking care to move only with the flood tide, and to anchor in deep water.

The banks for the first 30 miles were fringed with tussoc. Above that some trees appeared; and at the distance of 40 miles, the banks were skirted with a strip of dense forest, of moderately sized trees. We proceeded 10 miles further up, when the country appeared to be chiefly covered with large trees, wide apart, with no undergrowth between them, excepting on some low spots near the river, with here and there a spot of open prairie.

Having anchored the Explorer In a safe situation, I landed with a boat's crew at one of the open spaces, to examine the productions of the land, and see if I could discover any indications of inhabitants. I found the timber to be mostly different from that which I was acquainted with, excepting a species of fir resembling our spruce. I was much pleased to see wood of this description, and immediately ordered the launch on shore, with the axes and all our disposable force. We were busily engaged for three days in filling the Explorer with wood for fuel, and, after stowing her quite full, piled as much on deck as I thought she would bear, including timber for constructing winter quarters for the sealing party.

All fears of the consequences of wintering in this region were now done away. Where trees could live, I could live. I determined to erect a secure establishment for my sealing party and pursue my discoveries as far as practicable. While the woodcutting was going on, I did not venture far from the river— I had not forgotten the big tracks. I was always on shore with the party, to be ready for events, taking the people all on board with me when I wanted a four hours' nap.

I employed myself in searching for curiosities, collecting geological, mineralogical, and ornithological specimens, sea fowl and land birds being very numerous in this country, and in gathering plants to enrich my hortus siccus, for the benefit of the learned when I should return home. My research was rewarded by the discovery of some enormous bones, possibly of a whale, which being, according to very high authority, no fish, might at some former period have got on shore in this high latitude, after the fashion of the other visitants from the internal world. As they were very large, I called them mammoth bones of course, had them all carefully taken on board, and packed in boxes, as an invaluable acquisition to the scientific world.

On the third day a cry of terror called my attention. I saw the men all running for the boats and thought it best to follow their example. We had all got into the boats, and shoved off into deep water, before I could ascertain the cause of the alarm, when the appearance of an enormous animal on the ground we had left answered my inquiries. The huge beast walked to the edge of the water at a moderate pace and stopped to survey us new corners with great composure. I ordered Jack Whiffle, who was an excellent marksman, to give him a shot from a three-pounder, mounted in the bow of the launch, and at the same time gave him a volley of musketry. Whether the shot took effect or not, could not be discovered. He returned to the woods without haste or fright, and thus deprived me of the pleasure of securing his skin and skeleton, for the examination of the learned, and the benefit of Scudder's Museum.

There was nothing to be gained by a longer continuance in this river, and I felt no disposition to penetrate into forests, frequented by animals large enough to be called mammoth, a name which appears to be applicable to all big things. At this place, fifty miles from its mouth, the river was full a mile in width, and twenty fathoms deep at low tide. Taking into consideration its unusual magnitude and depth, and the large animal seen upon its bank, I named it Mammoth River.

We arrived at Take-in harbor next day. Mr. Boneto's party had been actively employed and had already secured seven thousand seal skins. I collected all my officers on board and acquainted them with such of my plans as I thought it prudent to disclose. The first was to land thirty of the crew at a group of islands which formed a snug harbor near the mouth of Mammoth River; to erect on one of the islands sufficient buildings to protect them from the severity of the winter, in case it should become necessary to remain there until another season, and large enough to contain a fair share of all the stores on board, in proportion to their numbers, so that they might fare as well as those who remained in the ship. I told them that I should proceed to the S. E. along the coast, to ascertain where was the best sealing ground to remove to when these Islands should be cleared of seals, and to discover whether the land extended a sufficient distance on the other side of the pole to open a passage for us to sail over the pole, and thus proceed to Canton by steering due north, which would save a great deal of time. This was all according to their notions of things; but I was well aware that when they would suppose we were sailing northward on the other side of the globe, we should in fact be sailing directly into it through the opening.

No objections were made to this plan, as it all seemed feasible enough. But I was at a loss as to the officers I should leave with this party. In exploring the internal regions, I should want all my best officers; and although Slim was an excellent sealer, it would not do to leave him with the command of the party, for I should be sure to find the men all ripe for mutiny on my return. I at last determined to give Mr. Boneto the charge of the establishment, with the boatswain to assist him; to keep Albicore, Slim, and Mackerel in the ship, and give Jack Whiffle the birth of acting boatswain.

We were a week briskly engaged in carrying these arrangements into effect. Extensive buildings were erected of stone and wood, having a centre room, to which no external air could gain access, without passing through the flue of a stove. The storerooms were detached from the dwelling, that the stores might he saved in case of fire. A covered way quite around all the buildings, as well as from one to the other, was constructed, and the whole covered four feet thick on the sides and roof with the bog of the tussoc, timber and stone being placed on it to keep it from being forced off by high winds.

Having thus prepared for the safety and comfort of my people, I gave Mr. Boneto written instructions how to proceed in all imaginable cases, but especially cautioned him against going on to the mainland, lest he should be destroyed by the mammoth animal.

Aware that there was a *possibility* that I might miscarry, and never get back to this place, I devoted a day to the performance of a necessary duty to my country, namely, taking possession of the country I had discovered, in the name and on behalf of the people of the United States of America. I first drew up a manifesto, setting forth, that I, Adam Seaborn, mariner, a citizen of the United States of America, did, on the 5th day of

November, Anno Domini one thousand eight hundred and seventeen, first see and discover this southern continent, a part of which was between 78° and 84° south latitude, and stretching to the N. W., S. E., and S. W., beyond my knowledge; which land having never before been seen by any civilized people, and having been occupied for the full term of eighteen days by citizens of the said United States, whether it should prove to be in possession of any other people or not, provided they were not *Christians*, was and of right ought to be the sole property of the said people of the United States, by right of discovery and occupancy, according to the usages of Christian nations.

Having completed this important paper, which I composed with great care, knowing that many wars had been waged for a less cause than a right to so valuable a continent, I had it engraved on a plate of sheathing copper, with a spread eagle at the top, and at the bottom a bank, with 100-dollar bills tumbling out of the doors and windows, to denote the amazing quantity and solidity of the wealth of my country. When it was completed by the blacksmith, who was something of a proficient in the fine arts, I went on shore with all the officers and men that could be spared from the ship, taking my music, two pieces of cannon, some wine for my officers, and plenty of grog for the men. We marched up the shore with great pomp, the music playing and colors flying, to a convenient spot, where I buried the copper plate, and rolled upon it as large a stone as the whole ship's company could move, and ordered the blacksmith to engrave upon it, in large deep letters, "Seaborn's Land. A. D. 1817."

A liberty pole was then erected on the spot, and the standard of the United States displayed upon it; all of which being accomplished, I ordered a salute to be fired of one gun for every State. "How many will that be, sir?" asked Mr. Boneto, adding, that they came so fast he could not keep the run of them. Slim said it was twenty-one. I objected to that number, as being the royal salute of Great-Britain, and settled the matter by telling them to fire away till they were tired of it and finish off with a few squibs for the half-made States. We completed the ceremony with a plenty of grog, and reiterated huzzas, as usual, and thus established the title of the United States to this newly discovered country, in the most incontestable manner, and strictly according to rule.

CHAPTER V.

The Author discovers the south extremity of Seaborn's Land, which he names Cape Worlds End. The compass becomes useless. —He states the manner in which he obviated the difficulty occasioned thereby.—He enters the internal world: describes the phenomena which occur.—Discovers Token Island.—Occurrences at that Island.

I proceeded along the coast to the S. S. E. November 21st, 1817, the sun's altitude corrected for refraction placed us in a more northern latitude than we had left, which my officers considered as evidence of our having passed the pole and made some progress northward, and they accordingly congratulated me on the occasion. I knew better and was perfectly aware that if the poles were open, of which I had no doubt, we must necessarily change our apparent latitude by observation very fast; and on turning the edge of the opening have a vertical sun, an equal division of day and night, and all the phenomena of the equator.

To be prepared for this untried region, I calculated all the changes of the apparent altitude of the sun in all degrees of declination, as they must necessarily occur, assuming the form of the earth to be at the openings as stated by Capt. Symmes in his sublime theory; and formed tables that I might be able at any time to ascertain the ship's place without difficulty or delay.

We had thus far found the land to trend S. S. E. and S. Soon after noon this day we reached a cape, from which the land turned short round to the W. N. W. and continued in that direction as far as could be seen from the mast head. This being apparently the most extreme southern land of the external world, I named it Worlds-end Cape. I felt no disposition to follow the coast to the N. W. although it might be found to turn again to the south. The most prudent course appeared to be to keep sight of the land, that we might certainly find our way back again to Mr. Boneto's station. But a roundabout way to the internal world was not in accordance with my impatient feelings; and yet the indulgence of my desire required that I should manage with great circumspection.

The compass was now of no manner of use; the card turned round and round on the slightest agitation of the box, and the needle pointed sometimes one way and sometimes another, changing its position every five minutes. I had frequently heard Slim muttering his apprehensions, and even Albicore said to me, 'I hope we shall not have any bad weather or lose sight of the land.' My best seamen appeared confounded at the loss of the compass, and a degree of alarm pervaded the whole ship's company. I had foreseen the difficulty that might take place when I proposed to leave the land, and to avoid it had placed Slim on the larboard watch with Albicore, by which arrangement the charge of my watch (the starboard) when I was off deck, devolved on Will Mackerel, assisted by Jack Whiffle. This was mortifying to Slim, but he was aware that he deserved it.

I kept near Cape World's End, taking its bearings in a variety of positions, for the ostensible purpose of ascertaining its exact position, until four o'clock, when the larboard watch went below. I saw that both Albicore and Slim turned in to get some sleep, and immediately ordered Mackerel to keep the vessel on a course corresponding to south, and to press with both steam and canvass to the utmost. The wind was about N. W., fresh and very steady, which served as a guide, the helmsman being directed to keep the wind four points on the quarter. We ran at the rate of 16 knots. I gave strict orders that Albicore and Slim should not be disturbed at the usual hour of calling the dog watch; and when they came on deck at 10 P. M. there was no land in sight. The sun to their astonishment was just setting in the bosom of the ocean: they stared at one another, and looked at me, but said nothing. They were perfectly bewildered; they knew not which way was north, south, east or west. Had they now undertaken to direct the course of the vessel; they would have been more likely to run from the land than towards it. Mackerel was delighted to see the sun set once more; it seemed like old times; and the weather had been for some days so hot that a little night was very desirable.

I told them all to be perfectly at ease, for that I knew what I was about; that I could calculate every point of the compass, as well as if that instrument performed its office; that we would have to for the night, the occurrence of which was no more than I had calculated on; and finally, to give them confidence in my skill, told them, that if we did not find the sun directly overhead at noon, within two days, provided no land impeded our progress, I would give up the command to Albicore, and show him the way back to Seaborn's Land.

Albicore and Slim both earnestly entreated that I would instruct them how to calculate the points of the compass, if I possessed that important knowledge, so that they might be enabled to find their way back again in case any accident should befall me. I begged to be excused, choosing to keep the staff in my own hands.

The truth was, having three excellent chronometers, one set to the time at Washington, one to that of Greenwich, and the other to that of Rio de Janeiro, and also an excellent watch daily regulated, which gave me the ship's diurnal time accurately, I could easily calculate my longitude, and the point on which the sun ought to hear every hour in the 24. With these calculations before me, I had but to look at my watch and the sun to determine my course. Thus, in the longitude of Greenwich, when the chronometer set to Greenwich time stood at 12 o'clock noon, wherever the sun was, was north; and when that chronometer stood at midnight, wherever the sun was, was south—on the external southern hemisphere, south of the degree of the sun's declination.

The reappearance of the stars, and the refreshing coolness of the night air delighted my people. At daylight we made sail and set the paddles in motion. At noon we had the sun nearly overhead, and the declination being 20° 5′ S. Slim was positive that we were in latitude 28° S. and wondered why the compass would not traverse. The next day we had a vertical sun, as I had predicted, and the weather as warm as I had ever known it at sea, with a fine breeze. No one knew which way we were steering but myself; and Slim's opinion confidently expressed that we were near the equator, and must soon make the continent of Asia, Africa, America, or the Asiatic islands, served to quiet the apprehensions of the men for their own safety, and at the same time to awaken their solicitude for the situation of Mr. Boneto's party, whom they said I had barbarously left to perish by the frosts of a polar winter, on Seaborn's Land.

The next day we observed the sun to the south of us, and nearly over head, and the compass began to traverse imperfectly. We had a regular recurrence of day and night, though the latter was very short, which I knew was occasioned by the rays of the sun being obstructed by the rim of the earth, when the external side of the part we were on turned towards the sun. The nights were not dark, when no clouds intervened to obstruct the rays of the sun, reflected from the opposite rim, and from a large luminous body northward, in the internal heavens, which reflected the sun as our moon does, and which I judged to be the second concentric sphere, according to Capt. Symmes. This gave us very pleasant nights, but not quite clear enough to render sailing through untried seas entirely safe.

We continued running due north, interned, three days, when the compass became pretty regular; but instead of the N. and S. points corresponding to the N. and S. points on the external world, as Capt. Symmes supposed it would do, the needle turned fairly end for end; the south end pointing directly into the globe towards the north pole, with some variation from the true north. But of this matter, I shall say very little, for sundry important reasons, and especially because intend to publish my theory of longitude in due season, and give the courses and bearings, corrected to true north and south, as understood by the externals.

On the 28th of November 1817, we discovered land, just at sunset, and immediately hove to, to keep a good offing until daylight. I walked the deck all night and was very impatient for the morning of that day which was to disclose to me the wonders of the internal world, and probably to decide the question whether it was or was not inhabited by rational beings.

Happily, day soon appeared, and we ran in with the land, keeping a good look-out, and the leads constantly going. On nearing the coast, we found the shore to be low and sandy. The body of the land, however, was high, with one towering peak far inland, Near the sea it appeared to be extremely barren, but some miles hack, scattered clumps of trees, and some appearances of verdure, afforded a more cheering prospect. We explored the coast of this island, for such it proved to be, for two days, before we found anchorage, or a safe landing place. A very heavy surf rolled on shore, and broke high on the shoals, which were frequent, and in some places three miles off the coast, so as to make it dangerous to approach. At length we found a safe road, sheltered by a sand bank above water, about two miles long, and lying parallel with the shore, half a league from it. There was a fair passage in, with 15 fathoms water, and good holding ground. Here we moored to the great joy of all on board, who, seeing firm land with living things of some kind moving about upon it, felt satisfied that they were still in the sublunary world, and complained of nothing but the excessive heat. It was near night when we came to anchor; all further research was therefore deferred until the next day.

On the 1st December, I landed for the first time on terra firma of the internal world, but was greatly disappointed, I must confess, to find no indications of any other inhabitants than turtles, terrapins of a monstrous size, some few seals, penguins, and numerous sea fowl. The great number of turtles was satisfactory evidence to my mind, that there were no human beings on the island; and, after a short walk on the burning sand, I returned on board, quite dejected.

The day was passed in fishing, and in collecting turtles and terrapins, for sea stock. In the evening, Mr. Slim, who was wide awake to his interest, suggested to me that we might obtain a good quantity of tortoise shell from this island, as the turtles brought on board were of the hawksbill kind, the shell of which sells for a high price. I gave him permission to land the following day, with ten men, and see what he could do in that way.

The next morning, I was quite sick, in consequence of the heat, and of my disappointment in not finding an inhabited country, after encountering so many hazards, and exerting so much enterprise and perseverance. Being thus compelled to remain on board, I permitted Albicore to land with four men, to ramble along shore, and see if he could make any discoveries. In the evening Slim reported that he had not been able to effect much, owing to the excessive heat, which compelled him, with his party, to take refuge under an awning, formed with the boats' sails, for full half the day. Albicore stated that he had been eight or ten miles along the shore but had seen nothing strange.

The following morning, when I had given orders to prepare for getting under weigh having determined to remain no longer in a place where there was great danger of the yellow fever making its appearance amongst my people, without intercourse with vessels from the West Indies, Albicore mentioned incidentally as we sat at breakfast, and as a matter of no sort of moment, that he had seen, during his walk on the beach, about five miles from where we lay, something which looked like part of a wreck of some outlandish vessel. The worthy man, who considered nothing that did not pertain to the strict line of his duty as deserving a thought, was astonished to see me spring up from my seat at table, order the boats manned, and make ready for an immediate expedition. It never occurred to his mind that if there were ships in those seas, there must also be men to build and sail them. To me the information he had given was both food and medicine: it revived my hopes and fired my curiosity. I felt no desire to complete my repast. I was restored to health and good spirits, and was soon marching over the sand, with Albicore for my guide.

After two hours we reached the place which Mr. Albicore had spoken of, where I found part of the frame of a vessel of some sort, of about one hundred tons burthen, the form of which satisfied me that it was no drift from the external world. The stern raked inwards, instead of out, as we construct them, giving the forward part of the vessel the form of a double ploughshare; while the broad bulging sides were admirably adapted to make the vessel sit firm on the water, and prevent her oversetting. But the most singular part was a piece of planking, which remained attached to the frame, and which was actually sewed on with a white elastic wire, resembling in appearance platina, more than any metal known to us. I extracted some small pieces of this singular metal, and with it fired the imagination of my people, by representing to them the enormous wealth we should acquire, could we obtain a cargo of it to carry to our country, where it would be more valuable than silver; and that the use to which it was applied was sufficient evidence of its being abundant where this vessel was built.

I named this island, which was in 81° 20´ internal south latitude, Token Island, considering its discovery as a token or premonition of some great things to come.

CHAPTER VI.

The Author departs from Token Island, in search of an internal continent. —Wind, weather, and other phenomena of the internal seas.—Great alarm of the crew.—Discovery of an inhabited country.

We were soon under weigh again, and steered due north, as well to seek for a new region of land, as to get into a more temperate climate; it being obvious that the internal equator must correspond in phenomena to the external pole, and consequently the more we approached the former, and receded from the latter, the cooler we should find the weather.

Soon after leaving the island, the weather became exceedingly unpleasant; the atmosphere was loaded with dense black clouds, and we were annoyed with torrents of rain, together with very vivid lightning and heavy thunder. We lay to the greater part of three days, thinking it imprudent to run into unexplored seas in dark weather. The fourth day it brightened up a little, when we pushed on to the northward.

After two days of unsettled weather, we were favored with a fine westerly wind, blowing steady and pleasant like a trade wind, which continued during the remainder of this passage. For three days more we continued steering to the northward, when we found the weather delightfully pleasant. We had the direct rays of the sun nearly one fourth part of the time, and its reflected light the remainder. This last was the most pleasant, being something between sunshine and bright moonlight, without the glare of the one or the indistinctness of the other. Satisfied with the climate, I determined to keep in it, and run before the wind due east, until I discovered land, or circumnavigated this part of the globe.

I found the latitude this day, carefully computed from the sun's altitude, with due allowance for refraction, to be 65° 17′ south internal. We ran on very pleasantly for seven days but saw nothing. It was now the 17th December. The sun had nearly attained its most southern declination and would soon be receding to the north.

The curious fact, that we could see the sun directly but for a short part of the day, at this season of the year, in a high southern latitude, astonished and alarmed my officers and people. It was a matter of continual debate amongst them on the forecastle, where Slim and even Albicore sometimes took a part in those grave and learned disquisitions. In one of their conferences, Slim advanced the opinion, that, as the sun was now near its extreme southern declination, and we could see it but a small part of the time, we must be in some great hole in the earth; and that when the sun returned to the north, which would soon take place, we should for a certainty be involved in total darkness, and never be able to find our way out again. This idea struck the whole ship's company with horror. Even Albicore was infected with the panic. Will Mackerel and Jack Whiffle were the only ones among them who expressed a ready determination to stand by their commander, wheresoever he might lead them. Numerous propositions were advanced and rejected by this council on the forecastle; but it was finally concluded that they would go aft in a body, and insist upon my immediately returning to Seaborn's Land, or they would heave me overboard, without further delay.

I was accordingly called from my cabin to hear this wise determination of my people. After hearing what they had to say, I asked them very coolly, how they intended to proceed when they had thrown me overboard? There was no one of them who could determine the ship's place, who had a sufficient knowledge of astronomy and natural philosophy, to account for the extraordinary phenomena that constantly occurred, or who had skill enough to ascertain any one point of the compass. How then were they to find their way home without my aid? Perceiving that this made a deep impression on their minds, I proceeded to dispel their fears, by assuring them that I felt no more disposition to perish in a sea of utter darkness than they did, but that so far from my having any apprehension of such an event, it appeared to me that we should find the winter in that region much more pleasant than at Seaborn's Land, if we could but discover land and a harbor, where we could moor in safety; that I had never been in a climate so perfectly agreeable to my feelings; that the air was so soft, so elastic, and temperate, it was a luxury to sit still and inhale the sweet breath of heaven; that so far from being in haste to get out of so salubrious a climate, I should be glad to pass my days in it; and, at all events, the sun would be no further north after the expiration of a month, than at the time of our departure from Boneto's station. Finally, I told them that, should I not make any discovery by the 1st of January, I would then return to Seaborn's Land, where, in the quarters erected for Mr. Boneto's party, we could all winter very comfortably; but, on the other hand, should they persist in their mutinous course, I would break my instruments, throw my books overboard. and leave them to help themselves as they could.

They all knew my determined and inflexible disposition, and that their best way was not to provoke it. The men went forward without reply. Albicore was the only one who opened his lips, and that was only to express his astonishment that he could have permitted himself to he led away from his duty for a moment, by any circumstance. It was all owing, he said, to that evil spirit, Slim, whose suggestion of total and perpetual darkness had frightened him.

We ran on for five days more, when "a sail ho!" rang through the ship. The stranger vessel was standing obliquely athwart our course, and we were soon near enough to see her distinctly from the deck. She had five masts, with narrow sails attached to each. When we were within three miles of the stranger, she tacked and stood from us to the southward, wind S. W. Feeling confident that the speed of my vessel was superior to that of anything on the face of the globe, inside or out, I gave chase, in expectation of bringing her to, in a short time.

But here I experienced a mortifying instance of the vanity of human pretensions, however well they may appear to be founded. The stranger, although she did not appear to have half as much sail in proportion to her hull as the Explorer, went within four points of the wind so rapidly, that in two hours she could not be seen from the mast head. I was now at a loss how to proceed. The strange sail was standing about N. W. when first seen, but she might be outward bound, and in that case, by steering that course we should miss the desired land; on the other hand, the course we had been steering might carry us to the northward of our object and pursuing the vessel in the direction in which she was last seen might lead to an equally unfortunate result. Will Mackerel was of opinion, that the Internals, on seeing so strange a looking vessel as ours, would run for the nearest land, and that we ought to follow her. I resolved at last to steer S. E. for two days, and if not successful, to return to the same place, and steer two days to the N. W. There proved to be no occasion for so much trouble; for at the moment, I had decided what to do, the lookout at the mast head called out 'land ho!'

The sun was now just setting, which immediately brought on the darkest period of the night; and some heavy black clouds occasioned by the vicinity of the land, threatened stormy weather. We therefore stood back upon our track, to wait the return of bright light, that I might approach the inhabited country of the Internal World for the first time under favor of the brightest smile of heaven.

After a few hours the clouds dispersed, and the reflected light became sufficiently strong to enable us to see dangers several miles, but not to admit of a clear distant view. We therefore drew slowly in with the land, to be ready to run into the nearest harbor during the next interval of sunshine. When near the shore, we again hove to with the ship's head offshore. With my night glass I could discern buildings and moving objects on the land, which assured me that the country was inhabited.

I walked the deck with impatient yet pleasing anxiety. I was about to reach the goal of all my wishes; to open an intercourse with a new world and with an unknown people; to unfold to the vein mortals of the external world new causes for admiration at the infinite diversity and excellence of the works of an inscrutable Deity; to give to them fresh motives for adoration and hopes of continued advancement in discovering the infinite works of God.

My imagination became fired with enthusiasm, and my heart elated with pride. I was about to secure to my name a conspicuous and imperishable place on the tablets of History, and a niche of the first order in the temple of Fame. I moved like one who trod on air; for whose achievements had equaled mine? The voyage of Columbus was but an excursion on a fishpond, and his discoveries, compared with mine, were but trifles; a summer sea and a strip of land, where common sense must have convinced any man of ordinary capacity that there must be land, unless Providence were in that one instance more wasteful of its works than in all its other doings. His was the discovery of a continent, mine of a new World!

My mind flew on the wings of thought to my native country; I compared my doings and my sensations with those of that swarm of sordid beings who waste their lives in Wall-street, or in the purlieus of the courts intent on gain, and scrambling for the wrecks of the property of their unfortunate fellow beings, or hiring out the efforts of their minds to perform such loathsome work as their employers would pay them for;—men who feel themselves ennobled by their wealth, or by their technical knowledge; who think themselves superior to the useful classes of society; from whom I had often heard the scornful observation, 'he is nothing but a shipmaster;' as if those men who live and thrive but by the infirmities and vices of society were ennobled by their profession, and the hardy and adventurous mariner, whose occupation leads him to every climate and through every sea, to gather like the bee the useful and the delicious for the comfort and gratification of the native hive, should be degraded by his calling.

CHAPTER VII.

Description of the first view of the coast. — The Author names the discovered country Symzonia. — Enters the harbour. — His first interview with the Symzonians. — Sketch of their appearance. — He commences the study of the Symzonian language.—Wonderful powers of mind displayed by the natives.—Account of an aerial vessel.

The mild oblique rays of the morning sun gilded to our view

"A scene surpassing Fancy's vision."

Gently rolling hills within an easy sloping shore, covered with verdure, checkered with groves of trees and shrubbery, studded with numerous white buildings, and animated with groups of men and cattle, all standing in relief near the foot of a lofty mountain, which in the distance reared its majestic head above the clouds, offered to mariners long confined to a wide waste of water the highest reward for their enterprise and perseverance;— the heartfelt satisfaction, that it was to their courage and skill that their fellow citizens would be indebted for the contemplation of so much loveliness. Here there was nothing wanting to a perfect landscape. Plain, hill, and dell sometimes rising with an easy slope, at others, broken, abrupt, or craggy; with an ocean in front, and a mountain in the rear, it was complete.

When the bright light of the sun first presented distant objects distinctly to our view, there were great numbers of vessels and boats in sight, mostly near the shore. We had repeatedly seen them during the night flitting past us like the shades of departed mortals. Immediately on observing our extraordinary appearance, they all retired towards an opening in the land to the northward, whither we followed them, and soon found that the apparent opening in the shore was occasioned by an island a short distance from the coast, having a roadstead within it, in which were several vessels at anchor. After hoisting out our boats, and seeing our guns in order, I stood in the roadstead, with my boats ahead. As we approached the anchorage, the vessels all retired into the mouth of a river which they ascended until quite out of sight.

At noon, on the 24th of December, we anchored in 14 fathoms water, on a fine sandy bottom. This land, out of gratitude to Capt. Spumes for his sublime theory, I immediately named Symzonia. The coast lay about S. S. W. and N. N. E. In the roadstead we were sheltered from all winds except those which blew directly along shore. These were not much to be feared, for we had found the prevailing W. S. W. winds to blow as steady as a trade wind for several days without any gales or stormy weather.

I passed an hour in surveying the enchanting scene by which I was surrounded; and in making preparations for a visit to the inhabitants of this internal world. I shaved my beard as smooth as I could, put on my best go-ashore clothes, and swung my hanger by my side, to make my appearance as imposing as possible. Here a difficulty occurred. I wanted an officer to leave in charge of the boat, on whose firmness and discretion I could rely in case of difficulty with the natives. I could not take Albicore, without leaving Slim in command of the Explorer, which was not to be thought of. I would not take Slim with me, for he would be more likely to contrive some way to get my throat cut out of sheer malice, than to use prudent measures for my safety. Will Mackerel was so hasty, that he would probably shoot the natives like pigeons, should he fancy them to be offering any offence or insult to his commander. I therefore determined to take Jack Whiffle, ostensibly to act as cockswain, with six of my best men, furnished with a musket, a pair of pistols, and a sabre each.

Thus equipped, and with the stripes and stars waving over the stern of the boat, I proceeded to the shore, having first instructed Albicore to offer no offence to any people who might approach the ship in my absence, unless it became necessary in actual self-defense, or to prevent them from taking possession of the vessel; and to inform me by signal should any superior force appear in the offing, or any danger be apprehended.

There were a number of buildings on the island, one of which from its magnitude and superior appearance to the others, I judged to be a public edifice of some sort. This structure was two stories high, while all the others were but one. In the front, a large open portico with an extensive platform, appeared to be a place of business, great numbers of people being collected upon it. In front of this building, a jettee into the water afforded convenient landing, and l directed the boat to be placed alongside of it. As I approached, all the people retired, and no sooner had I stepped upon the jettee than those in front of the large building moved into it.

Being determined to open an immediate communication with this people, who from the comforts with which they were surrounded could not be savages, I took off my sword, and gave it to Whiffle, and ordered him to lay off with the boat a half pistol shot from the shore, and not to fire a shot, nor to show his arms, unless he saw me run, or heard me fire a pistol; in which cases he must pull into the most convenient place to take me off, and to defend me.

I then walked slowly up the jettee. When I reached the head of it, I took off my hat and made a low bow towards the building, to show the Internals that I had some sense of politeness. No one appeared. I walked slowly up the sloping lawn, stopped, looked about me, and bowed, but still no one appeared to return my civilities. I walked on and had arrived within one hundred yards of the portico, when I recollected, that when Captain Ross was impeded in his progress northward by the northern 'icy hoop,' he met with some men on the ice who told him they came from the north, where there was land and an open sea. These men were swarthy, which Capt. Symmes attributes to their being inhabitants of the hot regions within the internal polar circle; in which opinion he was no doubt correct. I had frequently reflected on this circumstance, and had settled the matter in my mind that they were stragglers from the extreme north part of the internal regions; and could not but consider Capt. Ross as a very unfit person for an exploring expedition, or he would not have returned without ascertaining where those men came from, or how a great sea could exist to the northward of the 'icy hoop,' through fear of wintering in a climate where he saw men in existence who had passed all their lives there.

I remembered that these men so seen by Capt. Ross, saluted him by pulling their noses; and surely it is not surprising that men, inhabiting such different positions on this earth as the inside and outside of it, should differ so much as to consider that a compliment in the one place, which is deemed an insult in the other. Indeed, it seemed to me a small thing, when I considered how widely the most enlightened of the externals differ in opinion upon the most simple propositions of religion, politics, and political economy.

I was full in the faith that those men of Ross had been internals, and that their mode of salutation was much more likely to be in accordance with the manners of the Symzonians, than the rude fashion of us externals. I therefore pulled my nose very gracefully, without uncovering my head.

This was a happy thought. It arose from my having read much, seen a great deal of the world, and observed with tolerable accuracy, for a shipmaster, the important ceremonies and sublime rules of etiquette, by which the distinguished and the noble, the enlightened and the great, are implicitly governed; they being considered matters of more consequence than religious forms, or mere regulations of convenience.

I remembered that, on being honored with an audience of a sublime sovereign of the Mussulman empire, it was particularly enjoined upon me by the vizier, not to take my hat off; nor to sit cross-legged, the etiquette of the court forbidding any one to do so in the presence of the sovereign; and showing the top of the head or bottom of the feet being considered an insult to that exalted personage. Happily, I recalled to my mind all those weighty matters; and now, that I might not be guilty of insult to. this newfound people, I stood bolt upright, kept my hat on, and pulled my nose stoutly.

This had the desired effect. Several persons from within the building assembled on the platform of the portico. They stared much at me, which convinced me they were people of high fashion; conversed eagerly with one another and seemed undetermined how to act. More than one hundred men collected, before anyone showed any disposition to advance even to the front of the portico; and on the other hand, I dared not advance towards them, lest I should again put them all to flight, being already sensible that it was my dark and hideous appearance that created so much distrust amongst these beautiful natives. I therefore kept my position, occasionally pulling my nose out of politeness.

Full twenty minutes passed in this suspense; when one of the group, a man near five feet high, came to the threshold of the platform, and, raising his hand to his forehead, he brought it down to the point of his nose, and waved it gracefully in salutation, with a slight inclination of the body, but without actually pulling the nose as I had done, At the same time he spoke to me, in a soft, shrill, musical voice. His language was as unintelligible to me as the notes of a singing bird;. but his mode of salutation was not. I caught it with the aptness of a monkey, returned his courtesy after his own fashion, and answered him in English, with as soft a whine as I could affect, that my rude voice might not offend his ears.

Seeing him still in doubt whether it was a mortal or a goblin that stood before him, I bethought me to show him that I had some sense of a Supreme Being. I therefore fell on my knees, with my hands and eyes upraised to heaven, in the attitude of prayer. This was distinctly understood. It produced a shout of joy, which was followed by the immediate prostration of the whole party, who seemed absorbed in devotion for a few minutes. They then rose, and the one who had first advanced came towards me. I stood still to receive him, and as he walked close up to me, I extended my hand to ascertain if a thing so fair were tangible. He put out his hand and seized mine with a grip that made me start; but instantly let it go again and gazed upon me.

We spoke to each other in vain: he walked round and surveyed my person with eager curiosity. I did 'the like by him and had abundant cause; for the sootiest African does not differ more from us in darkness of skin and grossness of features, than this man did from me in fairness of complexion and delicacy of form. His arms were bare; his body was covered with a white garment, fitted to his shape, and hanging down to his knees. Upon his head he wore a tuft of feathers, curiously woven with his hair, which afforded shade to his forehead, and was a guard for his head against the rain. There was no appearance of any weapon about either him or any of the others.

Having both satisfied our eyes, I again endeavored to make myself intelligible to him; and, by the aid of signs, succeeded so far as to convince him that I came in peace, and meant no harm to anyone. He pointed to the building, which I took as an invitation to go in, and walked towards the portico, with the Internal by my side.

The fair skinned people by whom I was now surrounded, kept at a respectful distance from me. They formed a circle, and sat down upon their feet, with their bodies perfectly upright, and invited me to do the same. I admired the firmness of knee and strength of muscle which enabled them to make such a posture easy and pleasant, but took my seat on the floor cross-legged, like a Turk. Several of the principal men of the party seated themselves near me, and moved nearer and further off, as occasion required, with great facility, and without changing their sitting posture.

An amusing scene now occurred, while we endeavored to communicate our thoughts and wishes to one another. I shoved up the sleeve of my coat, to show them, by the inside of my arm, (which was always excluded from the sun,) that I was a white man. I am considered fair for an American, and my skin was always in my own country thought to be one of the finest and whitest. But when one of the internals placed his arm, always exposed to the weather, by the side of mine, the difference was truly mortifying. I was not a white man, compared with him.

I gave them to understand that I wanted food and drink, and immediately some delicious fruits, and a large bowl of excellent milk, were placed before me, which I ate with much satisfaction and an eager appetite, to the great amusement of the spectators, who seemed astonished at the enormous quantity I took. I afterwards learnt that what they set before me was sufficient for ten of these temperate beings.

The result of this interview was an understanding between us, that learning each other's languages would be the first essential step towards an intercourse between us; and for that purpose two persons were promptly singled out from the crowd, who took their seats by my side, with a writing apparatus, composed of some very delicate white leaves, more like sheets of very white ivory than like paper, and pencils which made a deep green mark.

We had scarcely entered on this important preliminary, when it was disagreeably and painfully interrupted by the firing of a gun on board the Explorer. The roar of a twelve-pounder, which jarred the building, struck a panic through the whole circle, and the volume of smoke which floated on the water alarmed them much.

With my pocket spyglass I observed that the signal for a fleet in the offing was flying, and it was to call my attention to this signal that Albicore had fired the gun. It was no easy matter to pacify the internals, and make them understand that the terrific noise, fire and smoke, were quite harmless. After many useless efforts, I made them comprehend that it was but the voice of the vessel, telling me it was time to return on hoard.

My spyglass attracted their notice. I gave it to one of the internals, and directed it to the ship, showing him how to find the focus. An exclamation of surprise showed me that this discovery in optics was unknown to them. This little incident was of great service to me. It showed the internals that some useful knowledge might be obtained from the hideous strangers and excited their curiosity to know more about us.

I now made signs to the two persons appointed to instruct me in the language, to accompany me on board, which after a few minutes they did, together with two others, ordered for the same purpose, and to make observations on our vessel and manners. I was the more willing to render this interview a short one, because I saw that no progress could be made until we had arranged some mode of communicating our ideas.

Soon after our arrival on board, a boat came off from the shore with a large supply of fruits and milk, which were most joyfully received; and in return, I sent a spyglass, a looking-glass, and several articles of glassware.

The vessels which had been signaled by Albicore, entered the bay and passed into the river, without coming any nearer to us than the land compelled them to. —After having shown my visitors about my ship, every part of which they examined with scrutinizing attention, I conducted them to my cabin, and sat down to the study of their language. Two devoted themselves to this object; the other two wrote an account of all they had observed and sent it by the boat which brought the fruits and milk.

I had not been long at my study of language, when Mr. Albicore sent me word that a bird as big as the ship was coming towards us. I went on deck, and immediately saw that Albicore's bird was no other than an ærial vessel, with a number of men on board. It came directly over the ship and descended so low that the people in it spoke with the internals who were with me; but I was not yet qualified to understand a word of what passed. I observed its appearance to be that of a ship's barge, with an inflated windsail, in the form of a cylinder, suspended longitudinally over it, leaving a space in which were the people. It had a rudder like a fish's tail, and fins or oars, which appeared to be moved by the people within. On the whole, it was not a matter of great surprise to me. I only inferred from it, that the internals understood ærostatics much better than the externals.

I afterwards learned that the air vessel over the boat was charged with an elastic gas, which was readily made by putting a small quantity of a very dense substance into some fluid, which disengaged a vast quantity of this light gas. By this means, the specific gravity of the vessel was diminished, in the same manner as that of a fish is by its sound. I also learned that this vessel had been dispatched by the government of the country to make observations upon the stranger who had entered their waters.

The following day I made preparations for another visit on shore, when I was made to understand by my instructors that I must not land again until I could speak the language of the country. I was not much pleased with this, not liking a confinement of two or three months, which, even with my faculty of learning languages, was the least term within which I could expect to qualify myself to speak one so new and difficult. My instructors, however, appeared very earnest on this point, and I thought it best to comply, and gave my undivided attention to the necessary study.

At the end of the first week, I was astonished and delighted to find my instructors addressing me in very good English. I could not help arguing, from their wonderful quickness of intellect, and faithfulness of memory, that I should find them intelligent and refined, beyond the conception of external mortals. In this I was not disappointed. My greatest misfortune was a want of capacity to comprehend intelligence so far beyond my powers of mind. They never forgot anything, and it was only necessary to name a thing once to fix it on their memories. The alphabet once read, and sounds pronounced, they had it perfectly, and expressed the greatest astonishment that I should require them to repeat the same names of things over five or six times, to fix them in my mind.

Having qualified themselves to act as interpreters, they acquainted me that permission had been given for me to visit the place of assembly, where the Best Man and the council of worthies were in session; but that my vessel must remain where she was, and none of the people be permitted to go out of her.

CHAPTER VIII.

The Author leaves the ship to visit the seat of government. — Description of the country.—Account of the polity of the Symzonians, as stated by his conductor.—Comparison of the industry, its objects and ends in the two worlds, and of the necessities and habits of the internals and externals.—Expulsion of the unworthy from Symzonia, to a place of exile near the north pole.—External world supposed to have been peopled by the outcasts.

It was the 2d of January 1818, that I set out on this delightful visit. A native boat came alongside to convey me, into which I stepped with no more sense of fear than might be excited on going among the spirits of the blessed; so perfectly did the appearance, manners, conduct, and expression of countenance of this people accord with my ideas of purity and goodness.

On the way to the place of assembly, which was about one hundred miles by water from the harbor where the ship lay, much occurred to gratify my senses, instruct my mind, and delight my heart. We ascended the river, the banks of which, and all the country near them, appeared like one beautiful and highly cultivated garden, with neat low buildings scattered throughout the scene. No crowded cities, the haunts of vice and misery, hung like wens upon the lovely face of nature. An appearance of equality in the condition and enjoyments of the people pervaded the country. The buildings were all of them large enough for comfort and convenience, but none of them so large, or so charged with ornament, as to appear to have been erected as monuments of the pride and folly of the proprietor.

Great numbers of small cattle and other domestic animals enriched the view, and a profusion of flowers, tastefully arranged in the vicinity of every house, filled the air with perfume, and charmed the eye with their variegated beauties. No fogs or vapors obscured the charming prospect, nor formed in windows to ornament the scene, the mild influence of the sun not being sufficient to produce rapid exhalations, nor the nights cold enough to condense them into vapor. Nature's fairest landscape requires no mantle to obscure its beauties, or to heighten their effect.

The active inhabitants all seemed engaged in something useful. Some were tending their cattle; some cultivating vegetables, fruits, and flowers, while others practiced the mechanic arts.

As we passed on through this enchanting country, Surui, the eldest of my conductors, instructed me in the civil polity, customs, mariners, and habits of this people. From him I learned. that in Symzonia all power emanated from the people; that the affairs of the nation were directed by

1. A chief, who was honored with the title of *Best Man*, and who held his situation for life, unless impeached of crime; but whose issue was considered ineligible to the same office for one generation after his decease.

2. An ordinary council of one hundred worthies, who assembled twice in each year, and oftener when circumstances made it necessary, to give advice to the Best Man.

3. A grand council of worthies, who assembled once in tour years, to admit members to their body, collect the sense of the nation on all public affairs, and aid the Best Man with their judgment in the appointment of Efficient to discharge the executive duties of the state.

The Best Man could only be elected by an unanimous vote of the grand council.

The Worthies are of three orders—the Good, the Wise, and the Useful.

The first, who have the title of *Good*, are such as have, by active benevolence, exemplary conduct, and constant efforts to promote the happiness of their fellow beings, obtained an expression of the public voice, that they are superior to the generality of men. When any such spontaneous testimony is given in favor of a man, it becomes the duty of the worthies of the district to which he belongs, to make the fact known to the grand council. The council examine minutely into the grounds of the popular opinion, and if they find it well founded, and that the man is truly good, benevolent, and virtuous, they admit him a member by the title of Good.

The second class of worthies are such as have in like manner been ascertained to have promoted the interests of society by improvements in science, and the advancement of useful knowledge. Such men, if free from vice, although not distinguished by benevolence, or the highest class of virtues, are admitted to the order of worthies by the title of *Wise*. This class corresponds to that of the philosophers of the external world.

The third, are all such as have manifested superior skill and diligence in their respective callings, with evident and constant good will towards their fellow men, such as have introduced useful inventions and improvements in the arts, set good examples to their neighbors, and are free from vicious propensities: these, on being found. justly entitled to such characters, are admitted to the order of worthies by the title of *Useful*.

The executive department is managed by Efficient, who are appointed by the Best Man, assisted by the Grand Council; and, in the interval of their session, if vacancies occur, by the ordinary council of One Hundred.

It is the duty of the worthies to notice the conduct of the people in their respective districts, to aid the feeble and distressed, if any such be found, to encourage the wavering, and reward the meritorious. Whenever any one of them discovers a man of retired but useful life, active but unobtrusive benevolence, extensive usefulness, with that modest shunning of the public exhibition of his doings which is necessary to possess the public in his favor, it becomes the duty of the Worthy to name him to the Grand Council, as a man of modest and exemplary merit; and if his character is, on investigation, found to be agreeable to the representation, he is admitted accordingly.

Amongst the standing rules of this body, Surui mentioned the following:

1st. Any man setting forth pretensions to superior merit, with a view to obtain a place among the worthies, is to be recorded as a *vain* man, and to be forever debarred the privilege of membership.

2d. Any man convicted of taking measures to gain a false reputation for merit, or of secretly influencing any person or persons to exert themselves to forward his nomination to the Grand Council, is to be recorded as a *deceitful* man, and to be thereby forever disqualified.

3d. Any man known to have been guilty of unjust or oppressive conduct towards any of those within his sphere of influence, or to have persecuted any who have been placed within the control of his power, to be recorded as a *tyrannical* man, and considered as wholly unfit to have any agency in the government of his fellow beings.

4th. Any man known to have affected a servile devotion to men of influence and power, or to have courted popularity by flattering the prejudices or passions of the people, to be recorded as a *hypocritical* man, and to be considered forever unworthy of admission to the distinguished orders.

5th. All persons guilty of crimes, all who infringe the rules of virtue and morality, all who lead irregular lives, or who set a bad example in society, are forever excluded from a place among the worthies. The last clause of this rule is understood to include old bachelors.

All the Efficients are appointed from the order of Worthies; no man being considered eligible to a place of trust who has not by his exemplary conduct, usefulness, and undeviating rectitude, acquired the notice and confidence of the public.

The Grand Council being very numerous, transacts business by a committee of three members from each district, to whom the other members communicate such information and advice as they may deem necessary. The recommendations to office are made by this body, to whom the cause of each nomination, and the qualifications of the persons nominated, are set forth in writing. The names of three individuals are always sent up to the Best Man, with a description of their qualifications and merits, for each office, of whom he selects and appoints the one who is, in his opinion, most deserving of it.

The exercise of intrigue and backstair influence being a bar to office, the offices of government are filled with the most intelligent, upright, and valuable men in the country, selected with the sole view of promoting the best interests of the nation.

I could not refrain from expressing my admiration of a system so wisely calculated to give the state the benefit of all the talents, information, and tried integrity of the nation. Surui asked with apparent surprise, if we the Externals did not select men to fill the places of honour, power, and trust, with the same scrupulous attention to their character, purity of life, usefulness in society, and goodness of heart. I was ashamed to acknowledge the truth and gave him a specimen of the veracity of an External by replying, "yes, much the same, at least in the State of New-York, where I am best acquainted."

I inquired whether the order of Worthies was a numerous body, and was informed that it embraced a majority of the men of mature age; all of whom were called in turn according to the order of their admission, to fill a place in the ordinary council of the Best Man. This council consists of one hundred; fifty-five of whom must be of the Good, forty of the Useful, and five of the Wise. Such persons, however, as had failed to maintain the character which obtained their admission to the order, might be excluded from a participation in the government; and the Best Man had the power to pass their names when summoning his ordinary council, provided he did not at any time pass more than one-tenth of the names on the list: for that proportion was deemed the utmost that could possibly deviate from the paths of rectitude.

It appeared to me to be a very troublesome form of government, which required the assemblage once in four years of more than half the men of the nation. But I found this to be a great mistake. Surui assured me, that the labor necessary to procure all the essential comforts and rational embellishments of life, in this fruitful country, and with the temperate habits of the people, required but a small proportion of the labor which could be performed; that there was abundant leisure for an annual assemblage of all the people, without any detriment to the business of society; and that every member of it enjoyed an abundance of the comforts of life, without excessive or constant labor. So far was the quadrennial assembly of the worthies from being regarded as an evil, that the arrival of the time of its occurrence was hailed as a season of great enjoyment, instruction, and usefulness.

The numerous inlets of the sea, which intersected this beautiful country in every direction, rendered travelling very easy and expeditions; so that not only the Worthies, but also such of their families as were of sufficient age to mingle with society, repaired to the district of the assembly, in which none but the Good, the Wise, and the Useful, were permitted to reside.—In the vessels in which they are conveyed, they take a sufficient quantity of substantial provisions for their own consumption, or to exchange for such as they may prefer during their visit. They also carry tents, in which such as cannot be accommodated in the houses of their friends reside, during their stay.

None but Worthies are permitted to enter the district of the assembly during the sitting. The first month of the assembly is passed in devotional exercises, and the interchange of visits and civilities, all vying with one another in endeavors to advance the happiness of those about them, and in conversing on matters important or useful to the commonweal. After comparing ideas with one another for a month, they appoint a committee from their number to sit in grand council one month, and no longer. All but the committee then return home, unless business, or a desire to offer their advice on some subject to the particular notice of the committee, induces them to remain.

Surui described the enjoyments of the season of the grand assemblage with the most enthusiastic expressions of delight.— None but the Good, the Wise, the Useful, none but the virtuous and benevolent, are then within the circuit of the district. The rarest gratifications of which the human mind is susceptible in intellectual intercourse, were then enjoyed without a sense of evil.

To me, who had been accustomed to see a great proportion of mankind constantly devoted to hard labor, or incessantly applying to business, to obtain a precarious subsistence; to see them, not content with the efforts which might be made by day, wearing out their health and lives in toil by the midnight lamp, and scarcely obtaining what are considered the necessaries of life,—it was difficult to comprehend how a great proportion of this people could leave their business and their homes, to pass months in a non-productive state, without oppressing the remainder of the people with intolerable burdens. But I was told that the Worthies received nothing for their services, and were able to provide without difficulty for themselves: all the revenue of the country was devoted to the maintenance of the Efficients, (who were paid for the time actually devoted to public affairs,) and to works of public utility.

This state of things appeared to me at first to be beyond the limits of possibility in the external world. My mind was for some time occupied by reflecting upon the extraordinary difference in the *natural* condition of the internals aid externals; and I commenced a comparison of the varieties and objects of industry in the two worlds, and of those necessities and habits which demanded the products of labor. This brought me to a clear view of the matter. I perceived that the greater part of the labor of the externals was devoted to the production of things useless or pernicious; and that of the things produced or acquired, the distribution, through defects in our social organization, was so unequal, that some few destroyed, without any increase of happiness to themselves, the products of the toil of multitudes. Instead of devoting our time to useful purposes, and living temperately on the wholesome gifts of Providence, like the blest internals, so as to preserve our health and strengthen our minds, thousands of us are employed in producing inebriating liquors, by the destruction of wholesome articles of food, to poison the bodies, enervate the minds, and corrupt the hearts, of our fellow beings. Other thousands waste their strength to procure stimulating weeds and narcotic substances from the extreme parts of the earth, for the purpose of exciting diseased appetites, whereby, in the case of those who possess good things, the ability to enjoy them is destroyed. Still greater numbers give their industry and their lives to the acquisition of mere matters of ornament, for the gratification of pride, an insatiable passion, which is only stimulated to increase its demands by every new indulgence. I saw that the internals owed their happiness to their rationality, to a conformity with the laws of nature and religion; and that the externals were miserable, from the indulgence of inordinate passions, and subjection to vicious propensities.

I inquired of Surui how I should know the distinguished orders? what badge or outward sign was worn by them? To which he replied, "They are known by their undeviating rectitude of conduct the good by their benevolence, the wise by their knowledge, the useful by their works." In answer to my inquiries as to the condition of those who were not of the order of Worthies, I was informed that it was very various, according to their conduct. Most of the people, seeing the happy condition of the Worthies, and being extremely desirous to partake of the refined enjoyments of the grand assemblage, strove earnestly to become deserving of a place among them; but some, giving way to their carnal appetites and passions, fell into intemperate indulgences, whereby they produced disease to their bodies, and a necessity for much labor to supply their unreasonable consumption, and at the same time an aversion to the performance of the labor which is necessary to the preservation of health; that the constant exhortations and efforts of the Worthy were found insufficient to restrain some of the youth from forming such pernicious habits, so that before they were sufficiently taught by experience and the examples before them, that to be good is to be happy, they degenerated into vice. This too often led to crime. To support their wastefulness, they infringed the rights of others. When such men became, in the opinions of the select worthies, incorrigible and dangerous to society, they were transported to a land far distant to the north, the extreme limit of the world, where a part of the year the heat is intense. There they continue in their vicious course, pursuing the gratification of their sensual appetites, and are punished with diseases of body which enervate their faculties, inordinate passions which torture their minds, and fierce desires which are incapable of being satisfied. —The influence of their gross appetites and of the climate, causes them to lose their fairness of complexion and beauty of form and feature. They become dark colored, ill favored, and misshapen men, not much superior to the brute creation. They retain, indeed, said Surui, some of the customs and manners of Symzonia; and the ceremony of pulling the nose in salutation by those who had strayed to the "icy hoop," and were seen by Captain Ross, of whom I had spoken, was no doubt a corruption of the graceful mode of salutation practised where I then was. On my first appearance, they had apprehended that I was of that outcast race; for it had been observed by those who had conveyed delinquents to the place of exile, that the descendants of the outcasts were enlarged in stature and size, owing to the grossness of their habits, and at the same time that they had lost their strength and activity. One of the pure race, it was believed, was able to lift three times as much as any one of the degenerates, or to leap three times as high. Their suspicions of my being of the outcast tribe, were allayed by the testimony of reverence to the Supreme Being which I had given, by falling on my knees, and imploring the aid of heaven in my embarrassed situation; whereby they knew that I could not be unworthy of their regard.

I felt not a little humbled by this account of the origin of the northern internal people, and cautiously avoided my observation that might discover, to my intelligent conductor, the suspicion which darted through my mind, that we the externals were indeed descendants of this exiled race; some of whom, penetrating the "icy hoop" near the continent of Asia or America, might have peopled the external world. The gross sensuality, intemperate passions, and beastly habits of the externals, all testified against us.

I inquired of Surui where this place of exile was situated. He said it was at the extreme northern part of the earth, as near the fountain of light and heat as mortals could go, without danger of perishing by fire: that they could only visit it in the temperate season, because during the rest of the year, the sun was seen directly overhead, when the heat was so great as to render existence extremely painful. By this account I knew that the place of exile must be situated somewhere on the verge of the rim of the north polar opening, as there, and there only, could the sun be seen directly overhead, without going to the external tropic.

CHAPTER IX.

The Author arrives at the seat of government. —Description of the Auditory. — Symzonian manner of assembling for devotion and public business. — Etiquette of the Symzonian Court. — He is admitted to an audience by the Best Man.—Account of the interview, and of his unfortunate efforts to exalt the character of the externals, by describing some of their splendid follies.

We were three days in passing from the ship to the place of assembly. Surui uniformly ordered a halt, when the light was so faint as not to permit me to have a distinct view of the country. Wherever we stopped, we were visited by great numbers of people, many of whom, to my extreme mortification, looked upon me with evident pity, if not with disgust. Yet they were very kind, and brought a profusion of the choicest fruits, vegetables, milk, and honey. and great quantities of beautiful flowers. —The face of the country became more and more beautiful as we approached the place of assemblage, which is in the most perfect part of this delightful region. The most elegant specimens of ornamental gardening on the external world, give but a faint idea of the appearance of this whole district.

The principal object that attracted my notice on arriving at the seat of government, was the Auditory, which towered above all surrounding objects, and struck me with awe and admiration. I could not conceive how so stupendous an edifice could have been reared by such a people. I had indeed observed, that notwithstanding their inferiority in size, they were much stronger, and more active than the Externals. The tallest men were about five feet high, but they leaped twenty or thirty feet at a bound without much apparent exertion, and easily lifted burthens which three of our men would find it difficult to move; still the vast fabric before me appeared out of all proportion to the ability even of mortals as highly gifted as these. It was a single dome of one arch, supported by a peristyle of huge columns, and covering at least eight acres of ground. The extreme elevation of the centre was seven hundred and fifty feet. The whole was formed of stone, in massy blocks, cemented with a paste of the same material, so as to appear to be all of one solid piece.

Surui said that when the people determined to erect a temple, in which they might assemble for devotional exercises and expressions of gratitude to the Divine Being, they regarded the object as one towards which the whole nation ought to be allowed to contribute. They determined to construct a building in which the greatest multitude ever collected in that district might worship God; and which would also serve for the deliberations of the Grand Council, that they might always be considered to be in the presence of the Supreme Ruler and discharge their high trust with a due sense of their responsibility to Him who seeth the heart. They had therefore built this Auditory by the surplus labor of the nation: each man having devoted so much of his time to the work as his private affairs would permit, and for no other reward than that of his own feelings and the good opinion of his fellow men.

The dome, which appeared so immense and so impracticable, was formed on a high conical hill, by which the site was originally occupied. In the sides of this hill shafts were sunk to the intended level of the foundation, in which the columns were reared. The top of the hill was then shaped for the reception of the stone of the arch, which was thus easily constructed upon the solid earth. When the whole was completed, the earth both within and without the structure was removed, leaving the edifice as it now appeared upon the plain. Within the columns, the earth was formed into a concavity, with graduated steps to the centre, so that an individual in any part of the immense area could see every person within the circumference of the dome.

In the centre, on a large convex platform, the Best Man has a seat, fixed upon a pivot, which permits him to turn with ease to every part of the Auditory. Over this platform an orchestra supported on pillars accommodates five hundred musicians, whose melody, reverberated by the vaulted roof, fills this tremendous and unbroken space.

In this edifice all the Worthies assemble once a day, for religious services, during the preparation month. The exercises are always commenced with music, to dispose the soul to heavenly contemplation. After the music, they all kneel and pray in silence. Speakers designated by the Best Man then ascend the platform by turns and deliver their sentiments on subjects worthy the attention of so enlightened and devout an assembly. The whole is closed with music, that all may depart in harmony of thought and feeling.

Three hours are thus devoted every day for a month, that the hearts and minds of the members may be improved, and that they may be prepared to deliberate upon the affairs of state in perfect fellowship and good will.

When the committee of the Grand Council, or the ordinary council of the Best Man, meet for the dispatch of business, they take their seats in compact order upon one side of the platform, leaving the area below for spectators; and as the most important matters are fully discussed in conversation during the preparation month, and as all the Worthies have good sense enough to know that their own happiness will be most certainly promoted by a faithful and pure devotion to the true interests of their fellow-men, the Best Man is not subjected to the inconvenience of listening for six hours together to a speech, calculated only to render a clear subject obscure and doubtful; nor is he who offers the fruits of his experience, or of his studies, insulted by the spectacle of an audience writing letters, reading newspapers, or sealing packets, to mark their contempt for his opinions.

I found a convenient and delightful lodge prepared for my reception. It was small, but sufficient for comfort. There were no servants attached to it, nor was there need of any. All necessary food, vegetables, fruits, milk, honey, &c. were sent daily, and placed where I could conveniently help myself. I soon learned that these supplies were voluntary contributions, and that the people took their turns in the privilege of administering to the wants of the stranger in their land.

Surui was accommodated in a similar manner, close by my dwelling. He passed a great part of the time with me; acting as linguist and continuing to teach me the language of the country, in which I was still very imperfect.

The day after my arrival, I was honored with an audience by the Best Man of this admirable people. I inquired of Surui as to the etiquette to be observed on going to court—whether I must uncover my head as in Europe, or my feet after the manner of the Asiatics? whether I must bow my head to the ground, making a right angle of my body, and walk backwards on retiring, as in the court of Great Britain, or flounder in flat on my belly, after the fashion of the Siamese? whether I was to stand or sit? if to sit, whether on the ground, or cross-legged, or on my haunches like a monkey?

Surui could not, or would not, understand me, and I concluded he wished the Best Man to see what the manners of an external would be, untaught in the customs of the country. I therefore determined to give them a specimen of the deportment of a republican freeman and conduct myself with the easy respectful politeness of a gentleman and citizen of the world.

On approaching the dwelling of the Best Man, I was charmed to find that it differed in no respect from the ordinary dwellings of the people, except that it was of greater extent, owing to his numerous families and a superior neatness and regularity was apparent in the grounds, which were stocked with a variety of the most beautiful and fragrant flowers and shrubbery. The house was literally in a bower of sweets.

The Best Man put me entirely at my ease in point of etiquette, by meeting me in the open air, in the garden, and without either the stiffness of affected pomp, or the austere visage of assumed sanctity. He received me with that frank, affectionate manner, which constitutes true politeness, the offspring of benevolence.

By the aid of Surui, we entered immediately into conversation. The first inquiries of the Best Man were, as to whence I came, and my motives for leaving my country. By means of a globe, which I had brought from the ship, and which I now caused to be produced, I explained to him the situation of my country, and the phenomena attending the external region, of which, till now, he had no conception, except from some supposed ravings of a *Wise* man, who was thought to be mad. The frightful glare of the sun, and the great extremes of heat, as his imagination pictured them in such an external region, were horrible to his apprehension.

My motive I stated to be, a desire to gain a more extended knowledge of the works of nature; adding, that I had undertaken this perilous voyage only to ascertain whether the body of this huge globe were an useless waste of sand and stones, contrary to the economy usually displayed in the works of Providence, or, according to the sublime conceptions of one of our *Wise* men, a series of concentric spheres, like a nest of boxes, inhabitable within and without, on every side, so as to accommodate the greatest possible number of intelligent beings.

I was already too well acquainted with the sentiments of this people, not to know that it would be extremely imprudent to suffer any expression to escape me which should discover that a desire of wealth; or of the means of sensual gratification, was among the motives which actuate the externals; such a disclosure being calculated only to excite their aversion and contempt.

The Best Man indulged me with a long interview; and it was a happy circumstance that I had with me a globe, charts, maps, books, and drawings, to illustrate and corroborate my statements; for otherwise I might have caused him to suspect that I was a most desperate liar, so strange and absurd did many of my representations appear to him. Happily, Surui was already able to read English books; and when I observed an appearance of doubt on the part of the Best Man, I sought out some passage in a printed work to corroborate my statement, which Surui translated into the language of the country.

I spoke of the danger I had encountered from ice. This was incomprehensible to him. He assured me that water never congealed in the internal world; that the *innate* warmth of the earth was sufficient to prevent it, and he could not understand how so great a degree of cold could exist in the external world, so much more exposed to the direct influence of the fountain of light and heat. I endeavored to. account for this by explaining to him the generation of cold by evaporation and absorption and promised to send to the ship for my air pump, to show him ice artificially produced by absorption in an exhausted receiver. I then proceeded to account for the equable heat in the internal world, and the extreme cold at the icy hoop, upon principles which appeared to me to be very obvious.

In the first place, the sun's direct influence is exerted on an equal portion of the globe at all times, which influence is felt, on the external surface, only where it is directly exerted. In such places it is felt intensely, but from the free action of the external atmosphere, so soon as that influence is withdrawn the heat escapes and flies off rapidly, generating cold in its passage, or by evaporation as we express it. Those parts of the external world from which the influence of the sun is withdrawn for the greatest length of time thus become intensely cold, excepting in the immediate vicinity of the polar openings, where the issue of warm air, from the internal cavity, tempers the atmosphere: but at a short distance from the verge of the opening the very influence of this warm air generates cold, by parting rapidly with its latent heat and condensing into snow and hail, which causes the circle of ice between the 70th and 80th degrees of latitude, called the "icy hoop." This escape of heat from the warm air which issues from the internal world, is so great as to irradiate the atmosphere near the polar openings; and in the extreme cold of winter, during the absence of the sun, this irradiation is so vivid as to be visible fifty degrees towards the equator, where the inhabitants, being fond of simple names, call it Aurora Borealis.

On the other hand, as an equal portion of the globe is at all times acted upon directly by the sun's rays, the internal contents of that globe must be always subject to the same degree of heat, excepting such variations as may be occasionally produced by the direct rays of the sun admitted through the polar openings. Of this fact we had evidence on the external world, where, in the most intense cold weather, we had but to penetrate a short distance into the earth to escape its influence. The temperature of mines, dug a short distance into the earth, was always above the freezing point; and the degree of heat at a given distance below the general surface of the earth, was found to be nearly the same in all latitudes, and at all seasons.

Hence what lie called the innate warmth of the earth, was nothing more than the collected heat of the sun absorbed and retained by the globe from the continued action of that luminary upon an equal proportion of it, at all times, in the same manner as a glass globe full of water, when set before a fire, will absorb and diffuse heat throughout the contents of the vessel equally, although but one side is exposed to the direct influence of the fire, while that part of the external surface of the vessel which is not exposed to the fire, but is subject to the influence of the cold air of the room, will obtain no other heat than may be communicated by the fluid within.

My printed books were subjects of great interest. The art of printing was unknown, although that of engraving was practiced. I explained the process of making and using types and promised the Best Man to instruct such persons as he might be pleased to direct, in the art, in return for the hospitality and civilities I had received.

He expressed a desire to be made acquainted with the form of government the religion, habits, sentiments and practices of the people of the external world, particularly as to our acquirements in useful knowledge: on all which subjects I was extremely disinclined to converse, being aware that if I spoke the truth I should fill him with disgust, and if I endeavored to disguise the truth, and to reply to his inquiries from my own imagination, I might be detected in falsehood, and deservedly turned with contempt out of the country.

To his inquiries respecting government, I replied by describing briefly the principles of the American constitution, taking care to say nothing about the qualifications for office, nor of the means resorted to obtain preferment. He thought the scheme well calculated for a very virtuous and enlightened people, but liable to many abuses through the want of a probationary course of qualification for places of trust and power.

On the subject of religion, I frankly confessed that every man was permitted to worship God according to the dictates of his own conscience, and that our government did not recognize one form of worship in preference to another. With this he appeared to be satisfied, but when I inadvertently added, that one of our wise men, who had filled the chair of "Best Man" of the nation, had expressed the opinion that it was of no importance whether the people worshipped one God or twenty, he started with horror, and expressed the greatest astonishment that an enlightened people should permit wise men to obtain controlling influence in a country; for, however useful and valuable they might be found to be within their proper sphere of action, like all powerful agents they were dangerous to the happiness of mankind if not restrained by powerful checks and controlling influences, to prevent their running into impracticable measures:—wherefore, not more than five wise men were permitted to sit in his council of one hundred.

On the subject of our habits, I was as brief as he would permit me to be and took especial care to speak only of the habits of the most virtuous, enlightened and truly refined people of our external world; but in spite of my caution, he extracted much from me which filled him with disgust and pity. That the most pure of our people should be afflicted with disease, was evidence to his mind that we were a contaminated race, descendants of a degenerated people. Having discovered from my remarks, that we ate the flesh of warm blooded animals, prepared in many forms with condiments and sauces to give it a higher relish, and, instead of confining ourselves to the pure fluid provided by nature to quench our thirst, that we indulged in fermented and distilled liquors even to inebriation, he was not at a loss for the cause of disease and misery, and was only surprised that such things were permitted, or, being permitted, that the race did not become extinct. Great inequality in the condition of our people, he inferred as a necessary consequence upon the indulgence in vice; because, while a virtuous man will perform so much of useful labor, or business of equal utility to society, as a matter of duty, as shall amount to his full share of consumption of the common stock of value, and if his labors be blessed with abundance, will not expend the surplus above his wants in things useless and pernicious or in the gratification of his passions, but bestow it upon the meritorious needy, to support the unfortunate, or in useful public works, the vicious man is rendered averse to the performance of his duty, and becomes wasteful of the products of the industry of others, without regarding the means, whether just or unjust, by which he may possess himself of them. Therefore, men feeding upon animal food and costly drinks, and given to the indulgence of inordinate passions, must of necessity become very unequal in their condition, depraved in their appetites, and miserable in proportion to their aberrations from the strictest temperance, virtue, and piety.

Finding that the longer we conversed on the ha hits, manners, and sentiments of the externals, the lower they would sink in the estimation of this truly enlightened man, I endeavored to turn the discourse to our acquisitions in useful knowledge, in full confidence that on this subject I should have a decided advantage and be able to raise the people of the external world to a high place in his consideration. I spoke of the perfection to which we had arrived in the manufacture of apparel; of muslins wrought so fine as not to obstruct the sight; and worth per square yard, the value of two months labor of an able-bodied man; of the shawls of Cashmere, so exquisitely made, as to be valued at two years labor of an industrious farmer or mechanic; of laces to ornament the dresses of our wives and daughters, one pound weight of which would amount to a sum sufficient to purchase the labour of four men for life; of splendid cut glass, and ornamental wares, dazzling to the eye of the beholder; of works of silver and of gold, so beautifully wrought, and so much valued, as to be objects of adoration to many of our people. The Best Man could hear me no further on this subject; he pronounced these things to be useless baubles, the creation of vanity, pernicious in their influence upon the foolish, who might be so weak as to place their affections on them, and the production of them a most preposterous perversion of the faculties bestowed upon us by a beneficent Creator for useful purposes. What possible use could there be for a garment, which would neither retain warmth to the body, nor protect it from external evils, or from the observation of others? And what apology could be found for wasting the labor of four men for life, which, properly directed, would supply the wants of twenty human beings, to provide ornaments for those who, if not arrayed in the white robes of purity and virtue, must be odious, although bedizened with all the finery which human ingenuity can devise.

I spoke of our skill in arms, in hopes to excite his admiration; of the invention of gunpowder; of fleets of ships for the transportation of armies to invade the countries of our enemies and contend in naval fight for the right of navigating the ocean. This was the most unhappy subject I had yet touched upon. Instead of exciting his admiration, I found it difficult to convince him that my account was true, for he could not conceive it possible that beings in outward form so much like himself, could be so entirely under the influence of base and diabolical passions, as to make a science of worrying and destroying each other, like the most detestable reptiles.

I felt a strong desire to draw directly from the fountain head of knowledge in this country, immediate information on a variety of subjects relating to the condition, sentiments, and knowledge of this remarkable people, but did not think it decorous to question so exalted a character in this my first interview. I therefore limited myself to a demand to be permitted to moor my vessel in a secure place in the river and remain until the return of the next summer's sun should render my return to the external world perfectly practicable.

I had but to explain the danger to which we should all be exposed, of perishing by cold on the passage, if I attempted to make it so late in the season, to obtain from the benevolent Best Man the desired permission to winter there; and orders were accordingly given to admit the Explorer into a river, and moor her in a place assigned for that purpose, but under an express stipulation that no person should land, or leave any communication with the people, unless officially authorized by the Best Man's orders, under the strict inspection of confidential Efficients. Enough had been already discovered of our sentiments and habits, to convince the Best Mari that a free communication with my people would endanger the morals and happiness of his.

To save myself the mortification of further conversation on the useful knowledge of the Externals, I promised to put all my books into the hands of Surui, to be translated into the language of the country; and having heard the Best Man's orders that every attention should be paid to my wants and those of my people, and that information on all subjects interesting to me, except the construction of their engine of defense, should be freely communicated to me, and the records of the assembly opened to my examination, I took my leave. The Best Man kindly ordered Surui to bring me often to his house, to converse on matters relating to the External World, and to the promotion of the happiness of our fellow beings.

CHAPTER X.

Containing some account of the strange rationality of the Symzonians. —Their simplicity of dress. —Manner of making cloth. — Circulating medium. —Taxes.

The friendly reception which had been given to me by the Best Man, and his commands that information should be freely communicated to the stranger, were a sufficient introduction for me to the notice and kind offices of these benevolent people. They required no other evidence that my rank was sufficiently elevated to render me a fit associate for them, than the fact that the Best Man had found my conversation so interesting as to induce him to pass several hours in my company. I was visited by all classes of the community and gave scope to my eager desire to possess myself of all the useful knowledge and science possessed by the most intelligent of the people.

I gave my attention in the first instance almost exclusively to the Wise, in expectation of finding their conversation most instructive; but I soon found that like our *philosophers* they were more given to abstract theories than to practical knowledge and would contend for hours to establish some fanciful hypothesis, to the neglect of plain and practical subjects of inquiry. I therefore turned my attention to the Good and the Useful, who never spoke on subjects they did not understand, and whose information, though not so abstruse as that of the Wise, extended to all matters of established utility. Moreover, they could be implicitly relied on; for having no favorite hypotheses to maintain, and no selfish ends to answer, they explained everything to me frankly and in an intelligible manner.

In this way, and by frequent interviews with the Best Man, as also by actual observation, I ascertained the following, among numberless other interesting facts:

That the fatal sin *cupidity*, which drove our first parents out of Paradise, is almost wholly unknown to the pure and uncontaminated Internals. They view the gifts of a bountiful Providence as an abundant supply of good things for the benefit of all, and sufficient to gratify all the rational wants of all the creatures for whom they are provided. They admire and adore the beneficence which could find pleasure in creating intelligent beings, and in providing for all their wants; and are emulous to approximate towards the spirit of love and goodness to which they are indebted for all their blessings. They are continually striving to improve themselves in this respect, by unceasing efforts to render one another, and all creatures within the sphere of their influence, happier and better; instead of exerting all their faculties, like the Externals, to gain advantages over their fellow men, to acquire the means of gratifying the worst passions of their nature, or to advance their own pleasures by rendering others miserable.

All the real wants of men in society are provided for in the most simple and natural manner. Usefulness is the test of value. That artificial wealth which exists amongst the Externals, and depends for its support upon their capricious passions, has no place with the Symzonians; our whole list of fancy articles, all our ornaments, every description of things which are only calculated to gratify pride or vanity, are considered by them as worse than useless. They wear garments because they defend the body and are necessary to decency; but it never occurred to their simple minds that the fairest work of an Infinite Being could be unproved by trinkets and fripperies of man's device. Their judgments are not so much perverted, nor their tastes so much depraved. Therefore, having ascertained a mode of providing necessary raiment in the most convenient manner, they one and all adopted it, and, by dressing alike, they maintain a perfect equality in their wants in that respect.

Their cloth is a beautiful substance, manufactured in a peculiar manner, by a process resembling that employed by the natives of the South Sea Islands, and not unlike our mode of making paper.

The material is found in caves and amongst the rocks of the mountains, where a species of insects, larger than our spider, produce it in great abundance. They form webs somewhat like those of spiders, but of a firmer texture, and more compactly woven. These webs have the properties of asbestos, owing probably to the insects subsisting upon that or some similar substance. The inhabitants collect them with great care and lay them in a mold of the dimensions of the piece of cloth to be made, placing so many of them one upon another as the intended thickness of the cloth requires. This done, a fluid preparation which hardens by the influence of fire, without losing its elasticity, is poured over it. It is then pressed firmly together, and passed over a heated cylinder, which completes the operation.

This cloth is extremely convenient. Being incombustible, like asbestos, it is only necessary to pass a garment through the fire to purify it perfectly. It is also very durable; and being exquisitely white, it corresponds admirably with the delicate complexions of the people, and the mild light of the region they inhabit.

All the divisions of labor necessary to the convenience and welfare of society, are here perfectly understood. The community is not bewildered by a voluminous and complex system of political economy, consisting of abstract principles, buried in abstract and unintelligible words, and rendered too intricate to be understood by those who have common sense, or too inapplicable to civilized society to be adopted by those who have any sort of sense—invented by the *Wise* men of one country to mislead the politicians of another, and to depress the Good and the Useful.

Their circulating medium consists of tokens for every variety of things, and every description of services. These tokens are originally issued by the government, for services performed and articles supplied for the national benefit. One description represents one day's labor; a second, a standard measure of grain; a third, a small measure of pulse; a fourth, a given quantity of a particular fruit; a fifth, a measure of cloth, and so forth. There being a sufficient variety to represent all the articles which are in common use, they have all the advantages of exchange, without the trouble of delivery when the things are not wanted for actual consumption.

When, by any circumstances, the supply of any particular article in any district falls short of the demand to such a degree, that the tokens will not command what they represent, it is the business of the government to draw from the more fruitful districts a sufficiency to equalize the value, either by direct purchase, or by requiring the contributions of the fruitful districts in kind, and sending the articles to the place of scarcity, or by receiving the contributions of the district in which scarcity prevails, in tokens, and thus raising their value, or by both these operations in extreme cases.

Commerce is practiced only for the common convenience of society. The accumulation of wealth, and indulgence in luxury, being disreputable, and a bar to admission to the distinguished orders, an overreaching and avaricious spirit is not generated by traffic, as in the external world, but every operation of trade and transfer is performed on the most reasonable terms, which will enable him who performs it to live upon an equality with his fellowmen.

All contributions are required directly from the people, that everyone may know the full extent of his proportion of the expense of government. Every man under the age of one hundred years, is rated at the same amount, unless he has young children; in which case the tax is reduced in proportion to the number of such children, according to a graduated scale. This tax is so light that nothing, but a criminal want of industry or frugality can hinder any one from paying it.

The whole revenue of government requires no more than one- or two-days labor of each man per annum; and as the government exists for the sole purpose of preserving the freedom of the citizens, in the pursuit of happiness, and in the enjoyment of all those privileges and immunities which are compatible with the well-being of society, all are equally indebted for its benefits. Property being altogether a matter of secondary consideration, is not considered a proper object of taxation. In case of an accumulation of good things in the hands of an individual, beyond his wants, the surplus is in general voluntarily devoted by him to the use and benefit of his fellow-beings, in some shape or other, for the promotion of his own happiness. Doing good is here considered as the highest of earthly gratifications. When a man is more than one hundred years of age. he is considered to have performed his full share of public service, and to be entitled to exemption for the remainder of his days.

CHAPTER XI.

Containing some account of the Symzonian engine of defense. — Story of a very ancient war with at; internal nation called Belzubians, which caused the invention of this engine. — Opposition of the Good men to its being used. —Fultria the inventor's speech in defense of it.—Deliberations of the Council.—Termination of the war.—Sentiments of the people on the subject.

The prohibition by the Best Man of any information being given to me of their engines of defense, excited my curiosity. I was anxious to discover what those engines were, and why a people so good, so benevolent, and so harmless, could have any occasion for them. I dared not ask for airy direct account of their nature, as I knew that an attempt to induce any one to break through the Best Man's injunctions, would be considered as the greatest of offences, and cause my immediate expulsion from the favored district.

All my efforts to obtain the desired information indirectly, fell short of my object. I however gathered from casual observation, and from some manuscripts which fell into my hands, that in times long past, the people of Symzonia maintained a commercial intercourse with a nation on the opposite side of the internal world, beyond the equator, called in their ancient manuscripts Belzubians. This intercourse was kept up for many years; for the *Wise* men contended that it was beneficial, by enabling the people of Symzonia to obtain many things *cheaper* than they could themselves produce them. But in process of time, the Good men discovered that the people became poorer, more addicted to idleness, and given to the indulgence of many inordinate desires and extravagant vanities. At length the unworthy became so numerous, as to endanger the morality and virtue of the whole community, and it was necessary to banish them to a place of exile. in the hot regions near the extreme limit of the world at the north.

This was the origin of the system of casting out the corrupted members of society. At the time of its adoption, the government endeavored to remove the cause of the evil by prohibiting, as they had a clear right to do, all further intercourse, for purposes of trade, with the Belzubians: but the latter had become so depraved and sordid, by their addiction to traffic, and were so puffed up with the idea that they were the most *powerful* nation of the two, that they resolved to maintain a commerce with the Symzonians by force, in defiance of the regulations of the government.

The Good men were thus placed in a most painful dilemma. They could not prevent this forced intercourse entirely, without shedding the blood of their fellow beings, to which they felt an insurmountable aversion. For a time, they contented themselves with endeavors to reclaim the people. They exhorted them to abstain from the use of the things brought by the Belzubians; and finally succeeded so far as to diminish the advantages which their enemies had before derived from the trade, so as to make it no longer worth pursuing. The Belzubians then sent armed men in their ships to take possession of Symzonia and compel the Good men and the people to submit to the contaminating intercourse demanded by their cupidity.

The most frightful distress now pervaded the land. The enemy having to do with a people who had no arms, and who were in the highest degree averse to the shedding of blood, easily conquered a large portion of the country. Those who had been corrupted by intercourse with them, joined the standard of the Belzubians, and forwarded their views.

The total subjection of the country, and the destruction of its virtues and happiness, would have ensued, but for the timely appearance amongst them of a man of singular ingenuity. This man, named Fultria, invented the air vessels, one of which I have before spoken of. He also invented the engine of defence, the description of which was prohibited. The knowledge of its construction, and the manner of using it, was confined to a few select Good men, who were bound to secrecy by the most solemn obligations. I could obtain no other idea of it, than that it was a vast machine moved upon wheels, and rendered of but little specific gravity, by means of the apparatus employed in their air vessels, by the help of which it could, on an emergency, be raised into the air for a short time, to cross rivers or broken ground.

It was propelled by means of a great number of tubes, projecting very obliquely through the bottom near the ground, through which air was forced with such prodigious violence, that the resistance of the earth and atmosphere impelled the machine forwards: in this way it was moved with astonishing velocity. From all sides of this engine a great number of double tubes projected, through which two kinds of gas were caused to issue. These gases uniting at the extremities, produced a flame of intense heat, like that of our compound blowpipe on a large scale, which flame, according to tradition, was ejected with such force, as to consume everything for half a mile in every direction. The interior of the machine was sufficiently capacious to admit men enough to direct its motions and prepare the gases, and also the materials and apparatus necessary to their production.

When this terrific engine was completed, Fultria proposed to exterminate the enemy at once; whereupon all the Good and many of the Wise men objected to so barbarous a proceeding. They contended that it would be contrary to the practice of civilized nations, and taking an unjustifiable advantage of the enemy, by using means of warfare not resorted to by civilized men, and not much better than poisoning them secretly. They could not consent to such unheard-of barbarity; at best it was justifying the means by the end and doing evil that good might come; but it was better to suffer wrong than to do wrong.

Fultria, on hearing these objections to the use of the means of emancipation which he had provided with so much labor and ingenuity, ascended the platform, and addressed the Best Man in Council in defense of his engine and his views. I found his speech on record, it is having been carefully preserved, notwithstanding the lapse of many centuries. I translated it with the aid of Surui, but our language is not sufficiently nervous to convey the sentiments of this enlightened man with the energy and conciseness of the original language. I endeavored to put down the substance of it in English, but it cannot be expected that a sailor should do such justice to a fine specimen of Symzonian eloquence as might be done by some of our professed belles letters scholars, who pass their lives in studying the arrangement of words and in admiring the elegance and dignity of their compositions.

FULTRIA'S SPEECH.

"Best Man of our race! you have been told that it would be barbarous in us to exterminate the corrupt and contaminating invaders—should we not be more barbarous to submit to depravement and degradation?

"You have been told that by using the engine I have invented, you would take an unjustifiable advantage of our foes—do they not take unjustifiable advantage by employing their superior skill in the diabolical arts of physical warfare and moral turpitude, to prostrate the strength and destroy the virtue of our people?

"You have heard it urged, that the practices of civilized men do not justify the use of such means of warfare, and that the adoption of them would be the extreme of inhumanity. What then shall we do? Shall we permit the wicked to gratify their cupidity by plundering the feeble and devastating the defenseless, with little danger of hardship to themselves, and many allurements of advantage from success, and thus perpetuate war by rendering the pursuit of it safe and attractive?

— "No, Sir; it is most humane to cut off the instigators and performers of inhuman deeds.

"I would show my abhorrence of war by rendering it too horrible to be encountered.

"I would abolish war by ensuring inevitable destruction to all who engaged in it.

"I would utterly destroy the invaders that none may hereafter dare to draw the sword for invasion.

"Let all who take the sword perish by the sword, and war will be known no more."

The Council deliberated upon the measures recommended by Fultria, and upon the miserable situation of the country. They had no support but their confidence in the Sovereign Ruler of the world, and no hope of relief but from the favor of his Providence. They feared that a majority of the people had now become so degenerate in their minds, and so exasperated by their circumstances, that they would be eager to second the views of Fultria and engage in the work of destruction. But, for themselves, with a few exceptions, they remained steadfast in their virtuous principles and feelings, and could by no means consent to do what every dictate of reason and religion forbade. They were accountable for their own acts, not for the acts of others, or their consequences. They knew that to do right, and that alone, was safe. If they acted, that must be their rule. The end could not justify the means.

At last, it was thought that the exhibition of this terrible machine, with all its engines in operation, in sight of the Belzubians and their adherents, would impress them with such dread and horror, as to drive them immediately from the country, and effectually deter them from ever returning. This expedient was therefore tried, and it was completely successful. The enemy fled with as much precipitancy and haste as did the Midianites at the sight of the lamps and the noise of the broken pitchers of Gideon. The land was presently cleared of the Belzubians and their apostate followers; all intercourse with their country was prohibited; and since that time war had not been known.

Three or four thousand years had now passed away, and doubts were entertained whether this were matter of genuine history, or an ingenious allegory, intended to present to the people a glowing picture of the evils which might follow a gross departure from purity of life and rectitude of principle. There were very few who could conceive it possible that human nature had ever sunk to such extreme depravity, and that so great a proportion of mankind had been enslaved by evil passions, as to render the wicked the most numerous. In general, therefore, it was supposed that Fultria, willing to exhibit a magnificent specimen of his genius, and being somewhat under the influence of vanity, as *Wise* men often are, fancied it possible for such a deplorable state of corruption and violence to happen in a long course of ages, and stated an imaginary case, as an excuse for constructing his tremendous engine.

I did not express my opinions on this subject, for I thought it most discreet to conceal the fact, that such a state of things actually existed in the external world.— My silence, however, did not avail; for, having put my books, among which were Ree's Cyclopediæ, Shakespeare's works, Milton's Paradise Lost, and many volumes of modern history, poetry, and novels, into the hands of Surui, I was soon called upon for explanations as to what was true, and what fictitious.

CHAPTER XII.

Wonderful faculties of the Symzonia. — Translation of my books into their language. —Proposition of a *Wise* man to make slaves of the Author and his people. —The Author's remonstrance. —The *Wise* man disgraced.

The extraordinary strength and vigor of the faculties of this people enabled them to effect, in a short time, what would occupy the most intelligent of the externals for years. I can convey an idea of them only, by calling to the recollection of the reader the talent at computation manifested by Zera Colburn, who, at the age of 10 years, calculated the sum of any given number of figures in the twinkling of an eye, as though he arrived at the result by intuition.

The faculties of the Symzonians all seemed to be nearly perfect. They are obviously as much superior to ours, as Colburn's powers of calculation were greater than those of other untaught boys; which, no doubt, results from their strict conformity to the law of their nature.

With such powers of mind, it need not be matter of surprise that all my books were very soon translated into their language, and numerous copies of them printed, and distributed amongst the most learned and discreet, with instructions to report as to their fitness for general circulation.

This examination and report brought me into serious difficulty. A certain *Wise* man presented a memorial to the Best Man, in council, in which he attempted to prove, from the books which I had put into his hands, my large size, dingy complexion, carnivorous appetite, and my own account of the sensual habits and propensities of my race, that we were actually the offspring of the wicked who had been expelled from Symzonia for their vices, and that we ought to be subjected to the penalty denounced by their laws in such cases.

On inquiry, I found the penalty alluded to in the *Wise* man's memorial, was nothing less than the delivering of such persons to the most severe of the *Useful* class, to be kept at hard work, poorly fed, and debarred from intercourse with the pure—in the hope that, in process or time, their gross appetites might be scourged out of them.

All the horrors of a rice swamp, with but a peck of corn a week for subsistence, sprang up in my affrighted imagination. I immediately set about an elaborate petition to the Best Man, in which I endeavored to refute the arguments advanced by the *Wise* man, and to show that my dingy complexion was owing to my seafaring life on the external world, whereby I was much sunburnt; and that the *Wise* man had been led into error by mistaking a work of imagination, for real history.

I admitted that there was a race on the external world, inhabitants of some islands far to the north, who, from their vicinity to the place of exile, might be the descendants of the outcasts, but who, in my opinion, were more probably the descendants of the Belzubians, being a restless, turbulent people, much given to depredations upon the rights and property of others, of insatiable ambition, inordinate avarice, and excessive vanity: who made war their chief occupation, maintaining vast fleets and armies; who plundered the feeble, enslaved the unwary, and levied contributions by force or fraud upon the whole human race:

That these islanders were a distinct people, who were regardless of the rights of others, being governed by cupidity, whereby they had become detestable to all the rest of the externals, and to my nation in particular, to so great a degree, that our *Wise* men (who have the control in the government, the *Good* and *Useful* being held in but little estimation by the wise and the useless in my country) had repeatedly ordained a non-intercourse, in the vain hope of bringing these supposed descendants of the Belzubians to a sense of justice; and that we were at this time only secure from their attacks, by an invention for blowing them into the air, if they ventured to assail our shores; that the book which had misled the *Wise* man was written by one of this people, and had no reference to my country.

Before I had completed my work to my satisfaction, I received the agreeable intelligence, that the Best Man, supported by all the Good and most of the Useful of his council, had ordered the name of the *Wise* man who made the proposition, to be erased from the list of Worthies, as a cruel monster, for seriously proposing the infliction, upon strangers who had voluntarily thrown themselves upon the hospitality of the country, of penalties enacted only to render the consequences of the return of the outcasts too frightful to be encountered by them.

This was the only unpleasant occurrence during my stay. The days flew on with astonishing rapidity, so agreeably were they passed. The Symzonians slept but about three hours in the four-and-twenty and considered me a very gross and sluggish being because I could not do without six hours sleep. With the exception of this short interval, every moment was occupied in conversation, study, observation, or amusement. Statistics, geography, botany, ærology, geology, mineralogy, zoology, ornithology, ichthyology, conchology, and entomology, in turn demanded and received my attention.

CHAPTER XIII.

Recreations of the Symzonia. —Wonderful provision of nature for supplying the internal world with light. — Character and employments of the women of Symzonia.

I visited the place of recreation, a neat plain rotunda, in the centre of an extensive flower garden, where the young people, the middle aged, and the old occasionally convened, to extend their knowledge of one another, interchange their thoughts by conversation, listen to the most exquisite music, and practice a variety of graceful and elegant exercises. Being all very fond of music, they all join in that, by turns, as in other performances. Sometimes a hundred instruments, and many hundreds of the most exquisite voices, filled the whole place with the most enchanting sounds.

The exquisite beauty of the women, the graceful dignity of the men, the chaste decorum and sincere politeness of all, charmed the mind, and delighted the heart. Here there were no temptations to vice by offers of seducing cordials, wines, agreeable decoctions, or other intoxicating drinks, as in our places of resort for recreation. The enjoyments of this refined people were intellectual and pure—not the debasing gratifications of animal passions and sensual appetites.

The soft reflected light of the sun, which was now no longer directly visible, gave a pleasing mellowness to the scene, that was inexpressibly agreeable, being about midway between a bright moonlight and clear sunshine. I had great cause to admire the wonderful provision of nature, by which the internal world enjoyed almost perpetual light, without being subject at any time to the scorching heat which oppresses the bodies and irritates the passions of the inhabitants of the external surface.

When the sun has great southern declination, it is seen directly through the opening at the south pole, a little above the horizon—this gives an interval of bright light; and as the rays of heat are more refrangible than those of light *, a sufficient degree of heat is experienced to ripen the most delicate fruits.

At this season, during night, the rays of the sun are reflected from the opposite rim of the polar opening and afford so much light as to render the stars invisible. The full moon is never seen at this period; for while the sun is in south declination, the moon fulls to the north of the equator, to give light to the north polar region, and the northern internal hemisphere.

March and September are the darkest months. Both the sun and full moon are then in the equator, and shine very obliquely by refraction, into both polar openings. Yet, by reflection from side to side, they afford a faint light quite to the internal equator, where two reflected suns and moons are dimly seen at the same time. This circumstance had led the internals to suppose that there were actually duplicates of those luminaries. Their situation, it should be considered, did not admit of such observations of the celestial bodies, as were necessary to correct that error.

During this season, the planets and stars of the southern hemisphere are visible, some directly, and others by reflection. These occasions great mistakes in their astronomical calculations, which they ascribe to the aberrations of the heavenly bodies. It never occurred to them that their field of vision was a limited internal concave sphere, and a great part of their firmament nothing but a reflection of the external heavens.

When the sun is in north declination, it is not seen at all to the south; but as it then shines into the north polar opening, its influence is felt at Symzonia by a repeated reflection, and being aided both by the powerful light of the moon, (which always fulls in high south declination, when the sun is near the northern tropic, and shines directly into the southern opening,) and by the direct and reflected light of the planets and stars of the southern hemisphere, gives light enough for all necessary purposes.

The women of Symzonia are not regarded as inferior in intellectual capacity, or moral worth, to the other sex. The female character is there respected, for the qualities of the female mind are developed and employed. Their personal beauty exceeds my powers of description. I can liken their complexion to nothing but alabaster slightly tinged with rose. Compared with them the fairest of our fair are dingy. This may not be readily credited by some of our beauties, but they have only to place themselves near the alabaster ornaments in their drawing rooms to realize the fact.

The domestic duties of the Symzonian women are very simple, pleasing, and easily performed. To prepare the frugal family meal requires no roasting heat, nor black array of pots, kettles, spits, and gridirons. The little culinary preparation which vegetables and fruits require; is neatly and conveniently done in silver vessels; for silver is abundant, and well adapted for utensils for household use. To arrange their basins of milk and honey and set out their baskets of fruit for a family united in esteem and love, is a pleasurable exercise.

The preparation of clothing for a people of such simple habits requires comparatively little labour. The garden occupies a portion of their time, but the greater part is devoted to the instruction of their children, the improvement of their own minds, religion, and social intercourse.

Their parterres are not designed for the idle gratification of the eye, but to support innumerable swarms of honey making insects; the Symzonians being as fond of the sweet which nature has provided as the Externals are of that which is wrung from the bloody sweat of slavery.

Symmetry in form, and elegance in arrangement, are much attended to by these people; they do not attempt to surpass nature in the creation of beauties but endeavor to heighten the enjoyment of what is placed before them and make a right use of whatever they possess.

Vessels of gold for domestic purposes are sometimes used by those who cannot easily procure silver. Gold is abundant in the beds of rivers near the mountains, but it is not esteemed, because of its softness and great weight. It is chiefly employed in the fastening of their vessels, in place of iron, which is very rare, and much valued for its strength, and fitness for all the purposes of agriculture and mechanics.

CHAPTER XIV.

The Author examines the records of the Assembly. —Grounds of proposal for admittance to the order of Worthies—Shell fish of Symzonia.—Great quantities of Pearls, and the use to which they are applied.

I was allowed free access to the records of the Assembly: and, having made such proficiency in the Symzonian language as to read it with facility, I derived much amusement and instruction from the various recommendations for admittance to the distinguished orders which had been stated to the Grand Council and placed on record during a long course of ages. These records were much too voluminous to admit of my reading them in course. I therefore contented myself with opening them at hazard and reading whatever chanced to present itself.

One man was proposed to be admitted to the order of Worthies by the title "Wise," because he had given evidence of superior imagination and ingenuity; he having fancied that he had discovered by studying the laws of *matter and motion*, that the Internals were inhabitants of the concave side of a hollow sphere; and, reasoning from analogy, that the convex or outer side of that sphere must be inhabited by a people enjoying a wider range of action, and more extended views of objects floating in unlimited space: that the suns, moons and stars, which they saw imperfectly by refraction and reflection, were only visible through a dense atmosphere in their world, but must of necessity be directly visible to the inhabitants of the External World in all their effulgence. He had written a book to explain his ingenious theory of an External World, in which he had endeavored to show by various calculations, that his extravagant hypothesis was not absolutely beyond the limit of possibility.

This man was not proposed as one designated by the popular voice but was named by a certain *Wise* man as one of retired habits and uncommon genius. The council unanimously rejected the application and passed a vote of censure on him for troubling them with the dreams of a maniac or an enthusiast. The members of the council were generally of opinion that to suppose the outside of such a world to be inhabited was as absurd as to suppose men to dwell on the outside of their houses.

Another man was proposed as *Wise*, for devising a scheme to relieve the government from the trouble of superintending the distribution of things useful, in order to preserve equality in the comforts of the people throughout the land; and from constant attention to the emission and withdrawal of tokens, to maintain their regular value, and insure their proper effect. His plan was to substitute in place of the tokens a system of promissory obligations, to be issued by an association of individuals who should be always bound to *redeem* them. This plan, he contended, would greatly facilitate exchanges, and contribute to the convenience of government.

His scheme was promptly condemned, as a device to cheat the people, by causing perpetual fluctuations in the nominal price of things; and he was recorded as a designing man, unfit to be of the order of Worthies.

Another was proposed for admission as *Wise*, for composing a code of written laws, and writing a book to prove that the adoption of his project of numerous and particular laws in writing would conduce to the welfare of society, by enabling everyone to know, with technical precision, what lie might and what he might not do.

This man's scheme, and the proposition founded upon it were both rejected. The council said, that as to all the matters embraced in this proposed system, public opinion, the established principles and habits of the people, the prevalent sense of rectitude and benevolence, had been and still was sufficient. Laws, if in accordance with these principles, could add nothing to their efficacy; and if inconsistent with them, they could not be enforced. The whole subject was at present plain; technical phrases would but darken and perplex it. Language was imperfect; words had different meanings; those who violated the spirit of these laws would contrive to evade the letter; the people would disagree in their judgments; the influence of public opinion would be destroyed; bad passions would be generated; more laws would be required; contest, disorder, and innumerable evils would be the consequence. The education and discipline to which the people were accustomed, the examples of the Good, the dictates of enlightened consciences, the sense of accountability to God, the simplicity, temperance, and practical piety of the people, —these formed the basis of good conduct, and upon these dependance might be safely placed.

The most frequent grounds of recommendation for the distinguished orders were regular and useful industry, temperate and exemplary lives, and constant endeavours to improve themselves and others.

Many were admitted for discoveries in botany, whereby the people were enabled to derive increased enjoyment from the vegetable world; many also became Worthies by advancing the knowledge of entomology and finding how to guard against the ravages of insects, and how to turn the efforts of the myriads of almost invisible beings to harmless or useful ends.

I observed nothing of the nature of animals in use amongst this people as food, except oysters and other testaceous creatures, which have so little visible animation as to be considered by the Symzonians on an equality with vegetables, and to be provided like them for the nourishment of a higher order of life. They were probably led to this conclusion, by the vast profusion of shellfish which abound in their waters. They are caught in astonishing quantities. The shells are employed in building, and to promote vegetation.

The pearls, which they afford in great abundance, and of large size, are used to glaze the walls of their apartments, being dissolved in a liquid, and laid on like paint. This process gives a smooth and elegant surface, like the inside of the pearl oyster-shell, which is inexpressibly delicate and agreeable in the soft light of this country, and at the same time renders the walls more durable.

I visited a maker of this pearl wash. My cupidity, I must confess, was greatly excited by the sight of large heaps of pearls, which would be of incalculable value in the external world. Even in the atmosphere of this pure region, I could not prevent my imagination from figuring the splendid palace, dashing equipage, and choice wines I should enjoy, and the unbounded respect and obsequious attention which would be paid to me by the great men of Gotham, on my return there with the enormous wealth which a cargo of these pearls would produce. I asked the workman for a specimen of the pearls, and he gave me a handful that were as large as peas, which I put in my pocket, intending to show them to the Best Man, as a sample of the article with which I should be glad to load my ship.

CHAPTER XV.

The Author is ordered to depart from Symzonia. —The Best Man's reasons for sending him away. —His ineffectual efforts to obtain a place of rendezvous for purposes of trade.

It was on my return from this visit to the pearl wash maker, that I received notice to wait upon the Best Man. I immediately repaired to his dwelling, with a light heart, in expectation of my usual intellectual feast from his conversation, little suspecting that this interview was to be the last. He received me with a mild solemnity of manner, which warned me that the interview was for some purpose of importance. He did not keep me in suspense, but in a kind and benevolent manner informed me that the Wise men, to whom the copies of my books had been given, had all made their reports, which, together with the accounts of those who had observed the habits of myself and people, and been in the most favorable situation to ascertain my sentiments, had been submitted to him in council; that he had taken full time to reflect on the subject, before he determined on the painful measure which his duty to his people imposed upon him:

That, from the evidence before him, it appeared that we were of a race who had either wholly fallen from virtue, or were at least very much under the influence of the worst passions of our nature; that a great proportion of the race were governed by an inveterate selfishness, that canker of the soul, which is wholly incompatible with ingenuous and affectionate good-will towards our fellow-beings; that we were given to the practice of injustice, violence, and oppression, even to such a degree as to maintain bodies of armed men, trained to destroy their fellow-creatures; that we were guilty of enslaving our fellow-men for the purpose of procuring the means of gratifying our sensual appetites; that we were inordinately addicted to traffic, and sent out our people to the extreme parts of the external world to procure, by exchange, or fraud, or force, things pernicious to the health and morals of those who receive them, and that this practice was carried so far as to be supported with armed ships, a thing unheard of, except from some very ancient manuscript accounts of the Belzubians, which had been considered by the Good men of Symzonia, for ages, as nothing more than fables.

After stating these and many other charges against the externals, he added, that many of his council seriously apprehended that it was only our inordinate thirst for gain, that had induced me and my people to hazard our lives in an unknown region, and that it had not escaped their notice, that my vessel was provided with terrific engines of destruction, no doubt to enforce our will where our purposes required it: Wherefore he, the Best Man, in council, had come to a resolution, that the safety and happiness of his people would be endangered by permitting any further intercourse with so corrupt and depraved a race. He therefore required that I should repair forthwith to my vessel. and there remain until the season of bright light was sufficiently advanced to enable me to return to my country in safety; and ordered that all necessary supplies of food, and whatever was wanted to refit my vessel, should be furnished at the expense of the state; but that I should not be permitted to take away any of the products of the country which I esteemed valuable for traffic, lest the cupidity of my countrymen should lead them to send an armed force to obtain such things.

They were fully aware, he said, of the articles which were most coveted by the externals; for my books had described them, and the purposes to which they were applied; and Efficients would therefore be appointed to examine my vessel and see that I took away none of those articles. He felt confident that they had additional security for a strict compliance with this prohibitory order, in my integrity, of which he had formed a favorable estimate, notwithstanding the corruption of my nature, and did not apprehend that I would break through his injunctions, after partaking so largely of the hospitality of the country.

I was petrified with confusion and shame, on hearing my race thus described as pestiferous beings, spreading moral disease and contamination by their intercourse, and by thus seeing all my hopes of unbounded wealth at once laid prostrate; and I did not recover from the despondency which overwhelmed me, till I recollected that Mr. Boneto would no doubt have a full cargo of seal skins ready against my return to Seaborn's Land, which would ensure me a handsome fortune.

Any attempt to dissuade the Best Man from his purpose, or to obtain a revocation of the decree, I knew would be altogether vain. I therefore endeavored to soften the judgment he had formed of the externals, by representing the books, from which the Wise men and himself had drawn their opinions, to be the works of the islanders whom I had described to him as the supposed descendants of the Belzubians, and that they were only re-printed in my country as they had been in his; that we professed to be much more enlightened than those islanders, and styled ourselves emphatically the most enlightened people on the face of the earth, by which we meant no disrespect to the Symzonians, the face of the earth being the outside of it only, and we were not sufficiently enlightened, when the declaration was made, to know that there existed any such people; and that there were many people amongst us who would eagerly emulate the purity and goodness of the Symzonians, could they but have the benefit of their example, and behold the happiness which attended their course of life. I specified one numerous class in particular, who were remarkable for simplicity of habits, active benevolence, and good will towards mankind.

I admitted that the permission of a free intercourse with the externals, might be productive of great mischief to his people, by introducing vice and disease, which had been observed to spring up amongst the South Sea islanders, and other unsophisticated nations, soon after their discovery by Europeans and Americans; but urged that a limited intercourse, under strict regulations, might be productive of much good; and that the Symzonians would, in that case, enjoy the sweet reflection, that they had contributed to the reformation of many of the externals, by the beauty and loveliness of their example, and at the same time have the benefit of more expanded views of the works of a beneficent Creator, through the information which they might derive from the externals.

To effect this very desirable end, I proposed, that Token Island should be established as a place of meeting and intercourse, where the externals might erect places of abode, and remain through the winter, and have communication with such of the Symzonians as the Best Man, in council, might be pleased to license for that purpose; and that the useful metal iron, which was not to be found in Symzonia in sufficient quantities to supply the wants of the people, was very abundant in the external world, and would be brought and exchanged for articles which the Symzonians considered useless, or nearly so.

The Best Man objected to this scheme. He had not forgotten the evils related to having followed the ancient commerce with the Belzubians. He also urged that Token Island was situated in the worst region of the earth, where the extreme heat and great humidity of the air would generate violent diseases amongst those who should have the temerity to remain there in the presence of the sun.

Unhappily, in my eagerness to carry my point, I assured him that this would be no objection to the externals; that in the pursuit of gain, they defied plague, pestilence, and famine; that the rich merchants who sent out adventurers, never took the climate of a country into consideration, viewing it as of no concern to them how many of the lives of shipmasters and mariners might be sacrificed, nor how many widows and orphans were thus created, provided they could make money by their business; that the externals would come to Token Island so long as there was anything to be gained by it, even if one half of their number should perish annually; and that the Symzonians could visit them in the temperate season, when they would be quite safe.

The Best Man heard me out, and then told me I had said enough. It would be much less dangerous to his people, he believed, to visit Token Island in the hottest season, than to hold intercourse with such a depraved, covetous, and sordid people at any time of the year. The plan was inadmissible—I must prepare for my departure—The decree would be rigidly enforced.

I expressed my reluctant acquiescence and begged to be fully informed of his will and pleasure, that I might not in any respect deviate from the course I was desired to pursue. I closed by expressing a hope that the numerous manuscripts which the Wise and the Good had bestowed upon me, might not be taken away, but that I might be permitted to carry them to my country, to instruct the externals in the wisdom they contained. After a moment's hesitation he replied that good books could not do harm in any world, and I might retain them. This was joyful to my ears. I felt sure of instructive and profitable employment for life in translating these productions for the benefit of my fellow externals, and took my leave of the Best Man, with the comfortable reflection that I had not discovered a new world wholly in vain.

On my return to my lodge, I found it deserted of the usual visitors, Surui and other Efficients, appointed to provide for my wants, being the only persons who approached or held conversation with me. —All other persons, from this time until my departure from Symzonia, avoided me in a manner as little calculated to hurt my feelings as possible.

My books were all returned to me; but, to mark my acquiescence in the justice and propriety of the measures adopted by the Best Man, I sent to him by Surui my best telescope, a solar microscope, an excellent sextant, a pair of globes, and a set of charts and maps of the external world. The instruments being superior to anything possessed by the Symzonians, and all these articles being calculated to extend their views of creation, I knew they would be highly esteemed. All these articles were cordially received as a tribute of gratitude on my part; and I was even given to understand that the Best Man derived more satisfaction from this indication of my heart, than from the possession of the very useful and desirable things I had presented to him.

CHAPTER XVI.

The Author returns to the Explorer—Holds a council of officers—Determines to return to Seaborn's Land—Takes leave of Surui—Sails from Symzonia—Touches at Token Island—Arrives at Boneto's station.

I returned to my ship, with sensations very different from those which delighted my heart on my passage from it. I felt like a culprit exiled to Botany Bay for his crimes: so strong was the contrast between the peaceful, intelligent, and virtuous people, from amongst whom I was driven, and the turbulent, rude, and corrupt externals, with whom I was doomed to pass the remainder of my days. My chief consolation was derived from that contemptible passion, vanity, a certain evidence that I was a true external. I could not avoid being elated, and indulging some pleasant emotions, when I thought of the great curiosity my arrival from the internal world would excite amongst the externals, the celebrity I should acquire, the prodigious importance which would be ascribed to my discoveries, and the unbounded encomiums which would be lavished on me for my wonderful capacity of mind, displayed in the contrivance of my voyage, and the incomparable bravery, skill, and perseverance displayed in the execution of it. All this, with the anticipation of the many public dinners which would be eaten in honor of the discoverer, the flattering toasts which would be drank all over the United States, and perhaps in Europe, together with the pleasure I should enjoy in relating my apparently tough stories, helped to keep up my spirits.

We were but ten hours in travelling to the ship; and it being the season of faint light, I could not make any new observations on the country. Surui and his companions were very reserved on the way. The little conversation which took place turned wholly on the beauty of holiness and purity of life, and the evidence of a blessed hereafter to all who are truly good.

I reached the ship on the 28th of July 1818 and found my people all very comfortable. Their chief complaint was that they had nothing fresh but oysters, which, in their opinion, were meagre food for civilized men, but which Mr. Albicore, to save our salt provisions, had given them very often. They found much fault also, that they were not permitted to go on shore. A profusion of the best of vegetables and fruits, with a full supply of the delicacies of the country, and with but little work to do, made them, as the like circumstances always make sailors, discontented and restless.

Surui having furnished me with a good chart of the internal seas, as far as Token Island, I determined to put to sea immediately, and proceed to that island, where I could employ my people in collecting tortoiseshell, until the sun should attain sufficient south declination to light our way back to Seaborn's Land. I accordingly called a council of officers, and laid before them the state of affairs, as far as I saw fit to disclose them, and the alternatives which were open to our selection.

In the first place, we might be able to find Belzubia, if we went in search of it, and if the people of that country retained their ancient habits, there would be no difficulty in opening a trade with them. On the other hand, if they continued to be a warlike and unjust people, they might have power and inclination to bale our vessel, and subject us all to slavery.

In the second place, Mr. Boneto's party would undoubtedly have a full cargo of seal skins ready for us against our return to Seaborn's Land, which would give us all money enough to make us comfortable at home; and it must not be forgotten that if we should go in search of Belzubia and be lost, Boneto and all his party must perish, and be lost to their country.

On the whole, I was willing to consider the discoveries I had made sufficient for one voyage, and to leave Belzubia for a subsequent expedition.

Slim's eyes glistened when I described the heaps of pearls I had seen, and he immediately proposed that we should possess ourselves of them by force, having no doubt that, with our firearms, we should be able to contend with any number of these delicate little beings, and thinking it of no manner of importance how many of them we might destroy, provided we got the pearls. But when I described to him their engines of defense, before which an army would disappear like a nest of caterpillars subjected to the flames of burning straw, his eyeballs swelled with fright, and he was anxious to put to sea with all practicable haste.

Albicore endeavored to account for the circumstance of the oysters sent on board having all been opened, and the soft part taken off by the Symzonians, by supposing that they did it to preserve the pearls for their own use; but it appeared to me to have been done because the impure part is not considered by them fit for food.

On the 13th of August, we put to sea. Surui accompanied us until we were quite out of sight of land, with a vessel in company to take him back. On parting with this excellent Symzonian, I presented him with a handsome gold watch, and a number of instruments and useful articles. He exhorted me to improve the instruction I had received while in his country, and to endeavor to imitate the morals and habits of the internals, as the only course by which I could advance my own happiness, and render myself better, and more capable of promoting the real welfare of my fellow-mortals. He also earnestly entreated me to warn my countrymen not to approach the coasts of Symzonia in expectation of being allowed any intercourse or traffic, whilst they remained besotted in vice and iniquity, the Best Man in council having decided, out of regard to the purity of the nation, that the engines of defense should be used to prevent such contamination.

We found no difficulty in making our passage to Token Island in twenty-one days, Surui having given me a particular account of the prevailing winds and currents, and the course to take to reach that island with the greatest expedition.

Here it may be well to explain the cause of the astonishing velocity of the Symzonian vessels, which enabled the one we had seen on approaching the coast to avoid us so easily. It appears that the Symzonians, in ancient times, apprehensive that the Belzubians might send armed ships to the coast to capture their vessels and carry away their people, devised a plan for accelerating their motion, by means of a number of tubes which perforated the after part of the vessel under water, through which air was forced with extreme violence by the agency of a curious engine, of which I could not obtain a particular description. This rush of air against the water forces the vessel forward with amazing rapidity. Every vessel going far from the coast must be furnished with one of these engines, but they are used only on emergencies.

The wreck of the vessel I had seen on Token Island was not of Symzonian construction, and the metal with which it was fastened was unknown in that country. It was the opinion of the Wise men that it must have been of Belzubian origin, for that people sometimes extended their voyages to Token Island to obtain turtles, which they eat.

On the passage to Token Island, I had very interesting employment in examining my Symzonian literary treasures, and in extracting and translating some of the most remarkable articles. The volumes which I had been permitted to bring away comprised a full account of all the science and useful knowledge of Symzonia; and in consequence of having this copious fountain to draw from, at pleasure, I had less occasion to depend on my written memoranda of the many curious and interesting facts and circumstances which fell under my observation whilst on shore.

These manuscripts were the only articles that I brought away from Symzonia, except the handful of pearls given to me by the pearl-wash maker, which, being concealed in my breeches pocket, and the fact of my possessing them being known only to the workman, I thought I might venture to smuggle, notwithstanding the Pest Man's confidential reliance on my integrity. This deviation from what was expected of me, will, I trust, be excused by my external friends, when they remember that I have been much addicted to commerce, and consider the force of habit, and the security with which the operation could be performed.

Soon after our arrival at Token Island, the sun was visible for a short interval at noon, nearly overhead. The remainder of the twenty-four hours we had a very bright light from the reflection of the sun and moon from the rim of the polar opening. Both those luminaries being now in the equator, their rays fell perpendicularly upon the rim of the opening, and being bent in by refraction, were visible at Token Island at noon. This direct and constant influence of the sun, the reflected rays being very powerful, rendered it very hot, early as it was in the season. Therefore, as I was eager and impatient to rejoin Mr. Boneto and his party, and to ascertain what success he had had in sealing, as well as to have several months of light, that my whole crew might, if necessary, be employed to complete a lading for the ship, we stayed but twenty days at Token Island. In this time we procured a considerable quantity of tortoise shell, and then proceeded direct for Seaborn's Land.

It was so early in the season that the temperature of the air changed rapidly, as we issued from the internal cavity, and approached the polar region of the external world. On the first of October, we experienced cold disagreeable weather, with slight falls of sleet and snow; but the sun was constantly above the horizon, and we pursued our course without delay. October 2d, we saw World's-end Cape, to the great joy of all on board, and especially of Mr. Slim, who could scarcely express his ecstasies. The following day we anchored in the harbor off Mr. Boneto's station, which, out of compliment to him, I named Boneto's Harbor.

We found the buildings, stores, and a large quantity of seal-skins carefully stacked, all in good condition, but no person on the island. I was immediately denominated a murderer, my men being certain that the whole party had been frozen to death, or that the mammoth animals had crossed on the ice during winter, and destroyed them all, so ready are seamen to put the worst construction on everything, and to censure their commander. Slim was "nothing loth" to forward this idea, a sight of the great quantity of valuable furs in which he was to share in no degree softening his malignity.

As there were no dead bodies, bones, or boats, to be seen, I was not alarmed for the safety of the men and had no doubt, but they were absent on a sealing excursion. The appearance of the boats under sail soon confirmed my opinion. We were presently joined by Mr. Boneto, who, with his party, had passed the winter very comfortably.

They had taken eighty thousand sealskins during our absence, most of which were preserved in salt, for the winter did not admit of their being cured by drying.

We had, therefore, abundant work before us to dry those skins, and to take a sufficient number in addition to complete our lading.

The joy of my officers and people, at this reunion, was without bounds. Sailors, on long voyages, become very much attached to one another, and consider every shipmate as a brother. I devoted three days to recreation, in consideration of the many perils we had encountered, and the great success which had thus far attended my enterprise.

CHAPTER XVII.

The Author loads the Explorer with seal skins, and sails from Seaborn's Land—Discovers Albicore's Islands. —Transactions at those islands. —He determines to conceal his discoveries from the world. —His reasons for this determination, and measures to effect it. —Sails for Canton.

We continued on the coast of Seaborn's Land until February 18th, when, having taken on board one hundred thousand seal skins, which were as many as we could stow without taking down our machinery, and that I did not think it prudent to do on that side of the icy hoop, we took our departure from Boneto's station, leaving all the animals that remained alive on one of the largest islands, to stock it for the benefit of future adventurers. We steered due north, and soon lost sight of the coast.

On the second day we fell in with extensive fields of ice, which compelled us to haul up, first N. W., then W. N. W., and at one time due west. This was somewhat alarming; but ultimately, we realized the correctness of my supposition, that the range of land must keep an open passage to leeward of it; and on the 1st of March 1819, I had the satisfaction to observe in latitude 69° 15′ south, with a clear open sea.

I now hauled up due east, to run down my longitude with the greater dispatch in this high latitude, where the degrees of longitude are small. This was fortunate: for by running on this parallel we discovered on the third day a group of small islands, forming a fine harbor, and well stocked with seal. Here we anchored. The islands were high broken rocks of granite and whinstone, apparently dislocated from their primitive bed, and thrown up by some volcanic eruption, or by the efforts of elastic gases generated in the 'mid-plane cavity,' to escape through this outer crust of the earth.

Some scanty tussoc, and a few mountain plants and mosses in the most favored spots, formed the only evidence of vegetation observable in these dreary islands. I named them Albicore's Islands, they having been first discovered by that vigilant officer; and determined to avail myself of the discovery to extend the profits of my voyage, by adding as many seal skins to my cargo as could be stowed in the space occupied by the steam engine and boiler, which I took to pieces, and placed in the bottom of the ship for ballast and dunnage. By caulking in the paddle ports, I also gained the place between the double sides, and rendered the ship to outward observation like an ordinary vessel, ketch rigged.

We remained at Albicore's Islands six weeks, in which time we obtained seventeen thousand skins. Having taken these on board and performed the important ceremony of taking possession of the islands for the United States, by hoisting the stripes and stars upon them in the usual manner, I was ready to depart. for Canton.

Being now about to visit a place where I should meet many of my countrymen and persons full of curiosity from every part of the world, who would be very inquisitive as to the discoveries I had made, I was led to reflect maturely on the consequences which might result from a disclosure of them; and the advantages which might be derived to myself, my friends, and my officers and people, by withholding all knowledge of them from the world.

At length, having made up my own mind on the subject, I called my officers and people together, and stated to them that if we should on our return to the United States, or at Canton, declare the discoveries we had made, we should in the first place expose ourselves to the charge of being impostors and outrageous falsifiers; in the second place, our countrymen, and even the Europeans, who would give us no credit for our bravery and enterprise, would avail themselves of all the information we might communicate, to fit out expeditions to Seaborn's Land, and possibly to Belzubia, and thus reap the harvest of our planting; but, worse than all, after thus appropriating to themselves the benefits of our skill and perseverance, they would assert that they had made all those discoveries, call all those places by new names, and affirm that we had never been there at all.

On the other hand, by concealing the knowledge of these discoveries in our own breasts, we could derive extensive benefits therefrom during the remainder of our lives. To effect this, they had only to bind themselves to me by oath, to keep this matter a. profound secret, and when they had been a sufficient time on shore, or had spent most of their money, I would fit out the Explorer, or another and better vessel, under the command of Mr. Boneto or Mr. Albicore, in which all should share according to their present standing on my books, and for which I should have money enough out of my profits from the present voyage. This would give us all a certain resource for the good things of this life; whereas if we made our adventures public, the business would be overdone in a year or two, and we should then have to look to the moon or some of the planets for room for further discoveries.

All assented to my proposal except Mr. Slim, who objected that an extra judicial oath would not be binding, that it would be a dereliction of duty on my part to withhold from mankind the knowledge of the most valuable part of the world, and finally, that he was principled against taking oaths. Slim was not open to persuasion. There was no moving him. He had gloated his imagination with the figure he should cut when, in consequence of having been an officer with me on this voyage, he should get command of a ship for a voyage to the sea of wealth, have the merchants crowding round him to obtain the benefit of the valuable information he possessed, and hear the delightful sound of 'Captain Slim.'

All my other officers and men took the prescribed obligation, whereby they bound themselves not to disclose by word, deed, writing, or sign, any of the discoveries or occurrences of this voyage, after our departure from off South Georgia, without my consent and approbation first obtained in writing. Slim's conduct was thought by all to be very unreasonable, and many of the men would willingly have thrown him overboard: but, with some difficulty, I pacified them, and persuaded them that Slim would think better of the matter before we reached Canton, if not, I would, while there, confine him to his stateroom, and prevent his doing mischief. in the hope that he would become more rational on the homeward passage. This important matter settled; we bore up for Canton.

CHAPTER XVIII.

The Author arrives at Canton. —Transactions in China, —Sails fur the United States. —Loss of manuscripts. — Difficulties with Mr. Slim.

We had a pleasant run to Macao Roads, with all the usual varieties of wind and weather. Having a full cargo of furs from the South Seas, a chop to proceed to Whampoa, the place where foreign ships unlade and lade their cargoes, was readily obtained. The Chinese regulations provide for the prompt admission of vessels actually laden with useful merchandise, but exclude all such as have no cargoes, which compels vessels that have nothing but ballast and specie to report their stores as cargo.

I received abundant civilities on my arrival at Canton. A shipmaster, with a cargo of three or four hundred thousand dollars at his disposal, is exposed to the most assiduous attentions. Upon this occasion my thanks were particularly merited by Mr. W. and Mr. C., both of whom very kindly proffered me all the services in their power for a moderate commission but in this, as in other instances, I preferred dealing directly with the natives, from the belief that they were quite as well versed in the business of their country as any foreigners could be.

Chien-loo, a native, obtained handsome offers for my cargo very promptly, and I soon sold the whole of my skins, large and small together, at two dollars and three quarters each. These, with the tortoise shell, produced the handsome sum of three hundred and thirty thousand dollars, clear of charges. I lost no time in selecting a cargo of teas, nankeens, and silks, and as much china ware as was necessary for dunnage.

Of the three hundred and thirty thousand dollars, one-third belonged to tiny officers and people, payable on their arrival in the United States, and two-thirds to myself as owner and master. Being rich, I now spent money freely, and advanced my officers and men as much as they wished to lay out; and after defraying port charges and other expenses, found I had a cargo of only three hundred and ten thousand dollars invoice; but, as the profits on this cargo were all to be my own, I reasonably calculated that on receiving them, I should be able to pay the balance due to the crew, and have a clear three hundred thousand dollars.

To make room for this cargo, I stowed the boxes containing the large bones, and my botanical, geological, mineralogical, zoological, ornithological, ichthyological, conchological, and entomological specimens, which were very extensive and valuable, in one of the paddle spaces between the double sides, and, to save a little room which remained, stowed a cable on top of them.

We touched at Angier Point, in the Island of Java, to fill up our water, and regale ourselves with the delicious mangosteens, which are there to be had in great perfection and abundance. That fruit is considered the most delicate and best flavored of any on the external world. Formerly it had given me great satisfaction; but now, after having enjoyed the exquisite fruits of Symzonia, it seemed quite insipid.

The day after leaving Angier Point, we were in the open ocean, with a stiff gale from S. E., driving us rapidly towards our homes, our wives and children. It is a delightful sensation which the mariner experiences on clearing port for his homeward passage, after a long and toilsome voyage. His home, his family, his little prattlers, and all the delightful associations of a happy fireside, crowd upon his imagination, which is cleared by long absence of all the asperities and disagreeable of real life. He flatters himself that he shall soon fold to his heart the wife of his bosom and the children of his love, improved in beauty, virtue, and affection; fancies a thousand enjoyments which the gains of his voyage will enable him to procure, and forgets the numberless vexations attendant upon business, and upon the duties of man in civilized society, encumbered with useless ceremonies and pernicious customs.

Mr. Slim had been confined to his stateroom whilst we lay at Whampoa; and no more intercourse was allowed between our people and their countrymen, than was necessary to keep up appearances. Our men were particularly cautioned not to drink grog whilst out of the ship, lest it should make them too talkative. They kept this injunction tolerably well for sailors; but one of them had nearly betrayed the whole secret, after drinking a second can of grog on board a Boston ship, where the Yankees seemed determined to get it all out of him. Happily, one of his shipmates forced him away, but not until enough had escaped him to produce a hundred absurd stories amongst the shipping in the river.

Being now at sea, Mr. Slim was permitted to go at large as usual. But alas! I had melancholy cause to regret this lenity. Having one day spread my Symzonian manuscripts on the after lockers, to dry away the mold which, from the humid atmosphere of the external world, had accumulated on them, I took a walk on the quarter-deck. On my return to my cabin, I was overwhelmed with consternation and alarm at the disappearance of my books and papers, which were all gone except my journal and volumes of extracts and translations. I immediately summoned the steward, but he could give no account of them. He had not been in my cabin during my absence. The cabin and stateroom were searched in vain. The manuscripts were gone! A man who had been working aloft, declared that he saw them going astern soon after

I came on deck and Will Mackerel, who was asleep in his birth, was positive that he saw the shadow of Slim passing from the direction of my cabin towards his stateroom. There was great cause to suspect that Slim had been into my cabin and thrown them all out of the windows to gratify his inveterate malice: but there was no help for it—there was no proof. A monkey, which, out of a foolish partiality to Jack Whiffle, I had permitted him to bring on board, and which visited every part of the ship, and was very mischievous, might have done it. They were irrevocably lost; and though I deplored them more than I should the loss of the mainmast, I was not without consolation. I had read most of them attentively, and being favored with a very retentive memory, I had treasured up their contents.

After this, I excluded Slim from my cabin, and kept a sharp eye upon him. Various modes were suggested by my officers and men, to obviate the difficulty which his refusal to accede to my measures threatened to produce. That which appeared most feasible, was, to confine him in irons, carry him home as a madman, and trust to the effect of his stories about the internal world, for a corroboration of his insanity. I however did not altogether like to trust to this maneuver, lest some of my people should prove treacherous, and, by joining their testimony to that of Slim, defeat all my projects.

My mind was suddenly diverted from this subject, which had long weighed heavily upon it, by the occurrence of real and immediate danger.

CHAPTER XIX.

Hurricane off the Isle of France. —Its consequences. —Death of Mr. Slim.

We were now to windward off the Isles of France and Bourbon, and nearly up with the land. This tract of ocean is the scene of the most violent hurricanes which are experienced on the external world, and it was our lot to encounter one of the most terrific.

A sudden change of the wind from S. E. to N. W. warned me of the coming storm. The ship was promptly secured for a gale; as much of the water which had been stowed on deck, was secured below, as the consumption of provisions had grade room for; the top gallant yards and masts were struck; booms sent down from the yards, dead lights secured, and every precaution taken to weather out the gale without damage. I never experienced a more awful tempest. The wind blew for some time with such violence as to make the face of the sea quite level, the pressure of the atmosphere, combined with its rapid motion, being so great as to prevent the swell from rising. The ship, under bare poles, drove broadside to the wind, nearly on her beam ends. When the violence of the first onset abated, the sea rose with a swell of full twenty feet perpendicular elevation. Having a strong vessel, although she was very deeply laden, I did not mind this much; but when the wind chopped round to the S. W., a heavy gale, bringing with it a large sea across the swell which the Northwester had produced, our situation was not devoid of danger. The tops of the waves, blown off by the wind, flew like the spray of a waterfall, and filled the air with water as high as the mast head while the waves, curled and lashed into foam by the whistling blast, gave the whole face of the ocean the appearance of one immense cataract. The vessel, assailed by the crossing sea from two points at once, labored excessively, and was fairly drowned with water. She frequently plunged the bowsprit quite out of sight beneath the wave, and had it not been of unusually firm construction, it must have gone to pieces.

Night set in without any abatement of the hurricane and served but to heighten the terror of its effects. The water in this part of the world, being charged with animalculæ or phosphoric matter, assumed in the darkness of the night, the appearance of a sea of liquid is boiling and whirling with ceaseless agitation. A poet would not need a better type from which to describe the infernal lake provided for the wicked.

Happily, we rode out the storm until nearly daylight, when the gale having abated, and there being every indication of more moderate weather, I went to my cabin to put on dry clothes, and left the deck in charge of Mr. Boneto, to whose watch Mr. Slim was now attached. I had not been long below when a violent shock, like that of a ship striking her side against a floating wreck, induced me to hasten back. I found my people in the greatest alarm, and the repeated blows, which made every timber in the ship tremble, were indeed sufficient cause of apprehension. I soon discovered the difficulty. The lashings of the starboard paddle port had given way; the port was open, and the shutter was swinging at liberty.

The gale had left a prodigious sea, which rolled the ship so much that at times she appeared to be going quite over. This caused the heavy port shutter, which was thirty feet long by three feet wide, to fly quite open, and then return against the side with frightful violence. It appeared that the lashings had been chafed in consequence of the boxes being badly stowed; and that the weight of the boxes in which were the large bones and all my scientific collections, together with the weight of the cable stowed upon the top of them, had burst open the port, through which the big bones, all my curiosities and ological treasures, as well as the cable, had launched into the sea!

To secure the port, which struck the ship with such force as to threaten to start the plink or fastenings, was an object of deep solicitude to everyone. Mr. Slim, for once, was very active and forward. He was evidently filled with apprehension of losing his life, or, what was not less dear to him, his share of the cargo; for, instead of looking deliberately about him to see what remedy was practicable, he seized a rope, and sprang into the space between the double sides, probably with the intention of fastening the shutter to the ring bolt, when it should swing to; but, losing his footing on the wet and slippery floor of the inner side, he launched half way out of the port, and as the ship rolled to windward, the slam of the shutter instantly killed him.

There was a sense of grief expressed in every countenance, on this melancholy occasion. Seamen invariably exhibit feeling for the sufferings and misfortunes of their comrades; however vicious and disagreeable they may have been.

The paddle port was, with great difficulty, secured, but without any other essential damage. Fine weather soon returned, and we pursued our course pleasantly towards home.

The remainder of the voyage was marked by no uncommon circumstance. When we approached the coast of America, I called my officers and men together; and endeavored to impress their minds with a strong sense of the importance of profound secrecy in relation to the subject of our voyage, and particularly enjoined upon them the necessity of refraining from liquor, which always makes sailors thoughtless and loquacious.

CHAPTER XX.

The Author arrives in the United States—Consigns his cargo to Mr. Slippery—Is reduced to poverty by the failure of Mr. Slippery. —His great distress. —Inducement to publish this brief account of his discoveries. —Conclusion.

On my arrival in port, I felt the importance which an ample fortune gives a man in this external world. The arrival of a South Sea ship from Canton, with a valuable China cargo, was no unusual occurrence, and excited no extraordinary interest; but it was speedily rumored that the Explorer had made a splendid voyage, and that Capt. Seaborn was as rich as a nabob. Abundant civilities were proffered to me, and numberless invitations to dinner were politely given.

I had now to select some merchant to assist in disposing of my cargo, my long absence, and consequent ignorance of dealers, rendering it imprudent for me to transact my own business; besides which, I found that, notwithstanding the whole of my merchandise was as much the product of American industry, as though I and my people had dug it out of the soil, (for instead of obtaining it with specie, we had procured it by our own manual labor,) I was required to pay or secure the enormous sum of one hundred and ninety thousand dollars duties to government. A strange thing surely, that the same tax should be levied on the privilege of bringing the fruits of our own industry into the country, as on cargoes bought with silver dollars, the carrying away of which impoverishes the nation. This did not seem altogether right either for individuals or the country; but there was no use in reasoning about it—it was required by law.

My own bonds for these duties could not be received, because I was not a permanent resident. In this exigency my friend, Mr. Worthy, occurred to my mind as a very fit man to act as my factor. He was an old acquaintance, a well-informed merchant, and a man of strict integrity; but, unhappily, at this time, rather low in credit, in consequence of having lost a great part of his capital by endorsing for his friends. It was doubtful whether his bonds would be thought sufficient at the Customhouse, and I was assured that he could not raise cash enough to answer the heavy demands which would be immediately made upon me by my crew, and my own expenses. Moreover, as I was now very rich, and had daughters nearly grown up, it was proper that I should gain a place in genteel society, whereas my friend Worthy, being a plain frugal citizen, did not mix with the *haut ton*, and could give me no assistance in that particular. All my friends (and they were now very numerous) protested against so foolish a step as that of putting all my affairs into his hands, for the sake of giving an honest man a commission of ten or twelve thousand dollars, when there were so many *great* merchants who would readily manage my concerns for a moderate percentage and introduce me to stylish society into the bargain.

I confess that the Symzonian doctrines had left so much impression on my mind, as to cause me some compunction at the thought of neglecting an opportunity to render my friend Worthy's family comfortable, by giving him my business, instead of bestowing the advantages of it upon a merchant rolling in wealth, who, after being roundly paid, would consider me under obligations for his services. My external habits and sentiments, however, got the better of my sympathies for my old friend, and, by the advice of my new friends, I addressed myself to Mr. Slippery.

Mr. Slippery was undoubtedly a great merchant. He lived in a spacious house in Broadway, rode in a splendid coach, walked like a man of consequence in Wall-street, was a bank director, and had the handsomest carpeted compting room in the city, and

I know not how many clerks writing in the next room. I knew him by sight, and did not altogether like to apply to him, because of his haughty manners. I remembered that when, some years before, I called at his compting room to offer myself as a master for one of his ships, he kept me standing half an hour, with my hat in my hand, before he condescended to notice me, and was no ways pleased that I took the liberty to draw a chair to seat myself until he might be at leisure. But he was certainly a great merchant, and to him I went.

I was delighted on entering his room, to observe a visible improvement in his deportment and manners. Instead of the distant, haughty reserve I had expected, he met me halfway, with both hands extended, and gave me a hearty welcome to my country after so long an absence; inquired after my wife and children in the most touching manner; was rejoiced to hear that I had made a great voyage and should be extremely happy to render me any service in his power. He finished his preliminary address with, "I am a great admirer, Captain Seaborn, of you men of enterprise, who draw riches from the great deep to the benefit of the revenue, the extension of trade. and all that sort of thing: you understand me, Sir?"

A hearty invitation to dinner, and a request to be permitted to introduce me to his friends, followed in a breath. I was charmed with him, poor fool that I was, little dreaming that it was the prospect of handling the half million of dollars, which my cargo would produce, that excited his cupidity.

There was no difficulty in settling terms. Mr. Slippery agreed to take charge of my business for half a commission, a simple two and an half per centum. He was aware, he said, that after a long voyage, I must be disposed to devote my time to my family and my friends, and he would take all the trouble of business off my hands. I had only to endorse over my bills of lading, and direct

Mr. Boneto to deliver the cargo to his order; and, as for money, I might draw for what sums I pleased, taking care, when I should draw for large amount; to make my bills at four or six months, as the goods must be sold on credit, and it would be a long time before be should be in funds from the actual proceeds.

A few months flew on delightfully; —I had no cares, no perplexities. Mr. Slippery recommended that the goods should be sold at auction, to make sure of the best of the endorsed paper, and I consented. He paid my officers and men their shares, as I desired; and although the auction sales did not produce for the goods, clear of charges, auction expenses, and Mr. Slippery's commission and guarantee, the actual cost in Canton, I flattered myself that I should still be rich enough, and at all events, I could send the Explorer on another voyage, whenever I should want more wealth. I purchased a handsome house for thirty thousand dollars, paid fifteen thousand dollars cash, and gave a mortgage for fifteen thousand; relieved the wants of all my poor relations; assisted many old acquaintances, who had been unfortunate; and still felt myself perfectly secure of all the good things of this world for the remainder of my days.

But alas! We are short-sighted creatures. I was soon called to lament the loss of my vessel, the partner of my adventures. Mr. Boneto not being satisfied with a life of idleness on shore, and having a wish to visit Europe, I permitted him to take the Explorer, without her machinery, for a voyage to New-Orleans, and thence to Europe. He took his money with him to purchase a cargo. On his way, he knocked that charming vessel to pieces on the Bahama Banks, for want of Blunt's chart, improved by recent surveys, to warn him of all the dangers.

This misfortune grieved me not only for my own loss, but for Boneto's, who was plundered by the Providence wreckers of every dollar. Yet it was but the beginning of affliction. A few days after, I was thunderstruck by a rumor that my friend the great merchant, Mr. Slippery, had stopped payment. But there was some comfort—I was assured that it was no failure, nothing but a suspension. For some time, I was kept at bay by promises and plausible statements. The whole truth, however, burst upon me at the appearance of Mr., Slippery's name in the Gazette, as an applicant for the benefit of the insolvent act.

My situation could no longer be concealed even from myself. I was utterly ruined. Many of my drafts on Mr. Slippery remained unpaid and came back upon me. I was sued and called a rascal for not paying my debts. No one would believe that the Nabob was actually poor. I pressed Mr. Slippery for assistance but got no other comfort than a cool recommendation to take the benefit of the act, as the most judicious course I could pursue.

I went to my family in a state bordering upon distraction. The troubles, mortifications, and miseries which followed, I forbear to dwell on. I endeavored to sell my house, but was told that property had depreciated so much, it was worth no more than the mortgage, for which the holder kindly took it off my hands. At length I was constrained to take Mr. Slippery's advice and apply for the benefit of the act abolishing imprisonment for debt.

I was now reduced to great straits, being *confined* to the *Liberties*, as they are called—for the enjoyment of which *restrained liberty* I found great difficulty in obtaining sufficient bail, my *friends* having entirely disappeared. Fortunately, I met with an old school-fellow, who, on hearing of my distress, proffered his bail, notwithstanding that the forfeiture of it would utterly ruin him.

At this period, when I frequently rose in the morning, without knowing how I should provide food for my children through the day, I found it difficult to feel and believe that it was all for the best. With neither the means of subsistence for my family, nor liberty to go in pursuit of them, my misfortunes and privations often weighed down my spirits, and became almost insupportable. When I thought of my situation, I felt no longer like a man. But the remembrance of the pious resignation, the humility, the contentment, the peacefulness and happiness of the Symzonians, recalled me to a conviction of the truth, that with a temper of calm and cordial submission to the will of Providence, a man may be happy under any circumstances, but without it must be wretched.

At this period of pecuniary distress, Will Mackerel accidentally heard of the misfortunes of his old commander, and hastened to see me. He could not comprehend why my being possessed of the *Liberties* should prevent me from going to sea, to acquire the means of subsistence for my family. The worthy fellow was wholly incompetent to understand the policy of depriving a man of liberty, preventing him from supplying the wants of those dependent on him, and compelling him to cast them as paupers upon the community, because he had, through misfortune, lost all his property.

Will had spent most of the money obtained by his voyage with me; but after hearing my story, and an account of the embarrassments under which I labored, he threw every dollar that remained to him upon the table and declared he would never touch a shilling of it whilst his old commander was in distress but would go to sea to render me further aid. I accepted this generous bounty with the frankness with which it was offered, and recorded Will in my heart as a true-hearted sailor. It was but little that he had left to bestow upon me, but it preserved me from the extremity of want for some time.

I was cheering myself with the prospect of obtaining my *real liberty*, and of persuading some man of capital to equip a suitable vessel for a second voyage to Seaborn's land, on terms which would give me a fair share of the advantages of the undertaking, when I was informed that Mr. Slippery had neither paid nor provided for the duties on the Explorer's cargo; that the bonds which he had given, owing to the long credits on China goods, were not yet due; and that, as I was the importer, I was responsible for the whole amount, and should be required to pay the uttermost farthing, or lie in jail during the pleasure of government, no insolvent act being considered of sufficient force to impair that *prerogative* of government, by which citizens were deprived of their liberty when misfortune had deprived them of everything else.

I had now no chance of freedom left, unless an opportunity should offer to fly the country before the bonds became due, for even should government relinquish the duties, the costs of suit, which amount in most cases to a large proportion of the debt, would not be relinquished till doomsday. To avail myself, however, of this only expedient, seemed impracticable. Even the shawls and trinkets which I had bestowed upon my wife in the days of our prosperity, were already sold, and the proceeds expended for bread. I was a fortnight in arrears to my landlady and had not a friend on earth from whom I could obtain a dollar. How then could I get away with nothing to pay my expenses, or those of my wife and children in my absence?

At this moment of difficulty, I heard that Captain Riley had obtained some pecuniary relief, by publishing a book of Travels, containing accounts not much more marvelous than those which I could relate of Symzonia. I therefore determined to make a brief extract from my journal for publication, to raise the wind, reserving most of the details of minute circumstances for my *personal narrative*, and my scientific research in statistics, geography, botany, aerology, geology, mineralogy, zoology, ornithology, ichthyology, conchology, entomology, horticulture, agriculture, to be digested hereafter under appropriate titles. The authenticity and genuineness of these research, since all the autographs and specimens collected to corroborate them were lost by the bursting open of the ship's paddle port, must rest upon the authority of my extracts, translations, journal, and memory. Should they even be questioned and disputed about by the *Scavans* of the external world, the generality of readers will probably trouble their heads very little on that score.

And now, kind reader, having transcribed thus much of my journal, in a manner which, I hope, will not be thought derogatory to the importance and dignity of the subject, I submit it to your inspection, with an intimation, that I am ready to undertake a second voyage to Seaborn's land, or a voyage to Belzubia and the place of exile, by the northern route, or another visit to Symzonia, and an ærial excursion thence to the inner spheres, as soon as I am furnished with the funds necessary to my escape from my present uncomfortable situation on the *Liberties*, in the garret of a lofty house, where, it being about the middle of dog-days, the sun exerts its utmost power upon the roof, within eighteen inches of my head.

www.ingramcontent.com/pod-product-compliance
Lightning Source LLC
Chambersburg PA
CBHW030621220526
45463CB00004B/1363